BIO-ART
to accompany

Asking About Life
TOBIN & DUSHECK

ALLAN TOBIN
University of California, Los Angeles

JENNIE DUSHECK
Santa Cruz, California

Saunders College Publishing
Harcourt Brace College Publishers

Fort Worth Philadelphia San Diego New York Orlando Austin
San Antonio Toronto Montreal London Sydney Tokyo

Printed in the United States of America

Saunders College Publishing: BioArt to accompany *Asking About Life, First Edition.* Tobin & Dusheck.

ISBN 0-03-023441-7

789 021 7654321

A Note to the Instructor

Thank you for adopting *Asking About Life* by Tobin & Dusheck. We appreciate your use of our text, and as a special aid for students we have produced Bio-Art™.

Bio-Art™ is a collection of important pieces of art from *Asking About Life* rendered in black and white. Generally most pieces of Bio-Art™ do not include labels, so students can use the art as a learning tool. It is an excellent tool in measuring student understanding of processes and organisms.

This valuable study aid is free with copyright privileges to instructors who adopt *Asking About Life* for class, or it can be purchased by students at a low cost. Please contact your bookstore if you would like Bio-Art™ to be purchased by students as a required or recommended supplement.

Suggested uses of Bio-Art™, for instructors who adopt *Asking About Life,* include:

- Copying Bio-Art™ as handouts for students, enabling students to label parts of figures, take notes, and avoid redrawing complicated diagrams, while the instructor uses a color overhead transparency from *Asking About Life.*

- Duplicating Bio-Art™ on overhead acetates and referring to and writing on the acetates during lectures, while students refer to their own handouts.

- Distributing copies of Bio-Art™ as part of exams and quizzes.

- Having students complete homework assignments using their own copies of Bio-Art™.

Asking About Life
Tobin/Dusheck

Bio-Art Figures

Figure 1.3   Biceps and Triceps

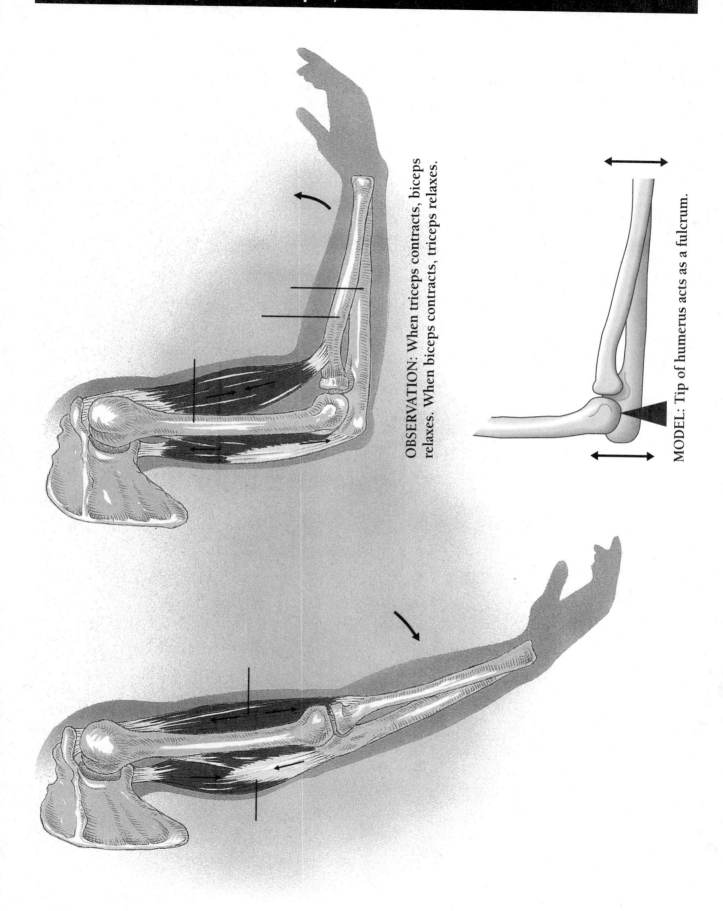

OBSERVATION: When triceps contracts, biceps relaxes. When biceps contracts, triceps relaxes.

MODEL: Tip of humerus acts as a fulcrum.

Figure 4.6 (L)   Plant Cell

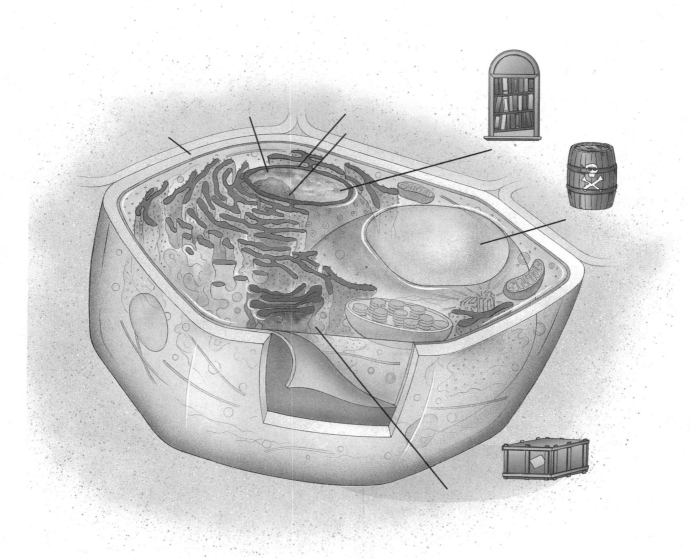

Figure 4.6 (R)   Animal Cell

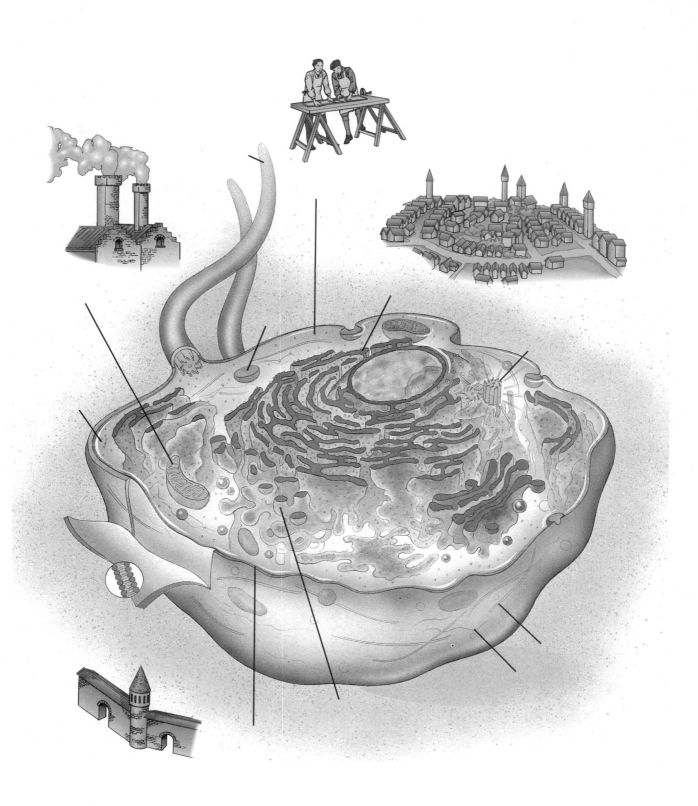

Figure 4.11b   Protein Transport

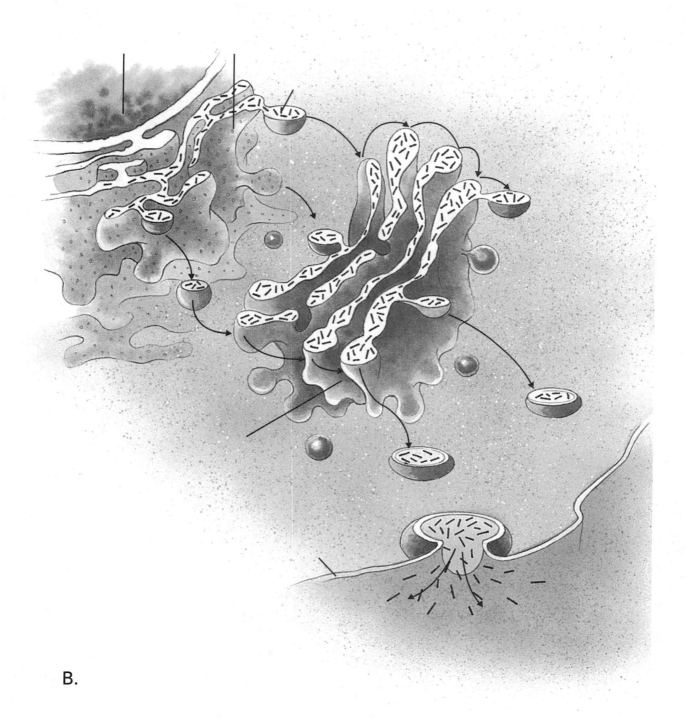

B.

Figure 4.14b Mitochondria

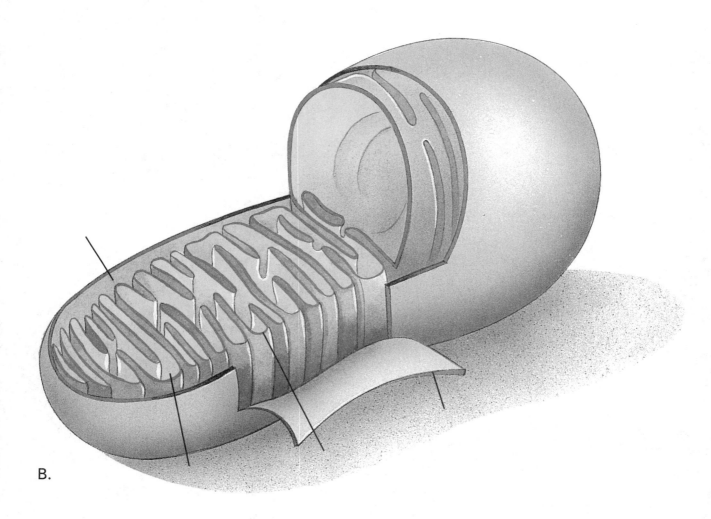

B.

Figure 4.15 Plastids

Figure 4.18   Plasma Membrane

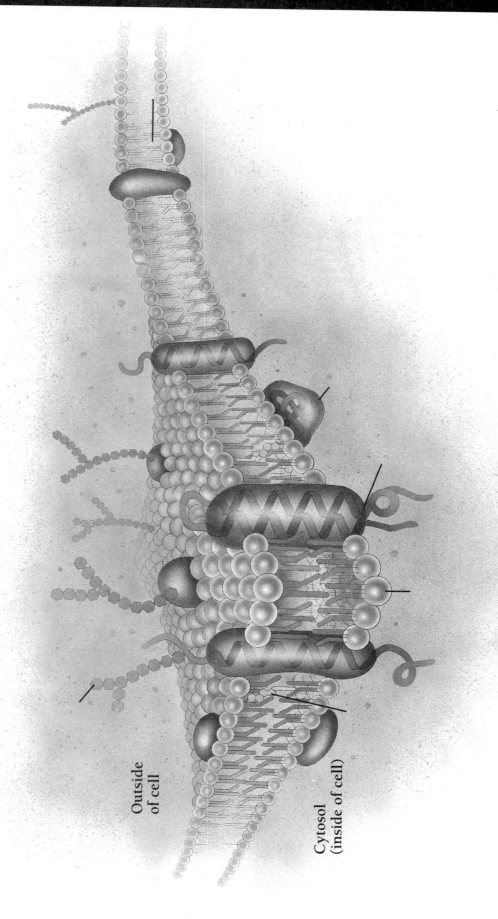

Outside
of cell

Cytosol
(inside of cell)

Figure 6.4   Cellular Respiration

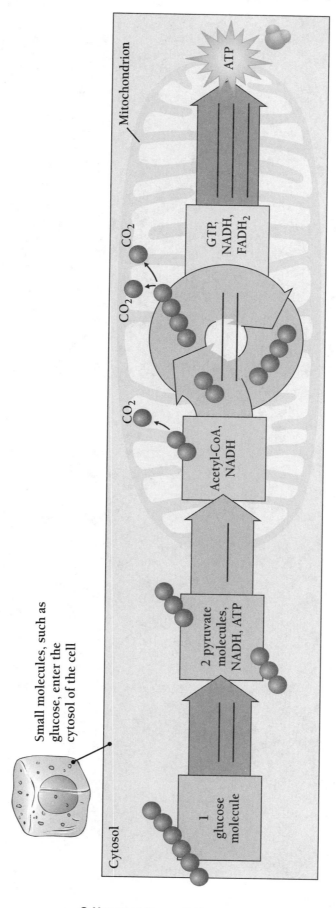

Mitochondrion

Cytosol

Small molecules, such as
glucose, enter the
cytosol of the cell

1
glucose
molecule

2 pyruvate
molecules,
NADH, ATP

Acetyl-CoA,
NADH

CO_2

CO_2

CO_2

GTP, NADH,
$FADH_2$

ATP

Figure 6.7   Pumping Protons

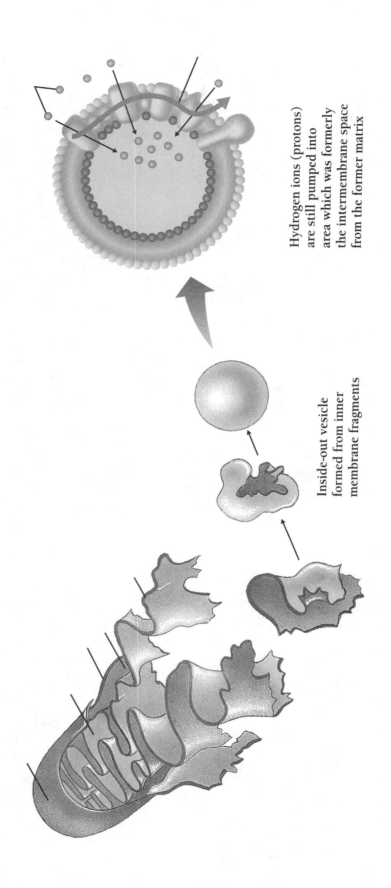

Hydrogen ions (protons) are still pumped into area which was formerly the intermembrane space from the former matrix

Inside-out vesicle formed from inner membrane fragments

Figure 7.14   Anatomy of a Chloroplast

Figure 7.14 (*Continued*)

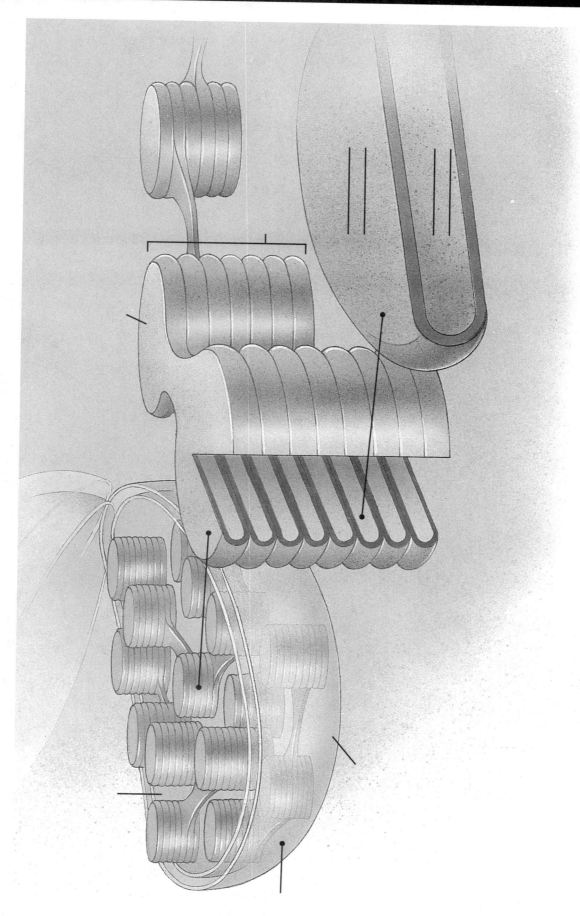

Figure 8.7   The Cell Cycle

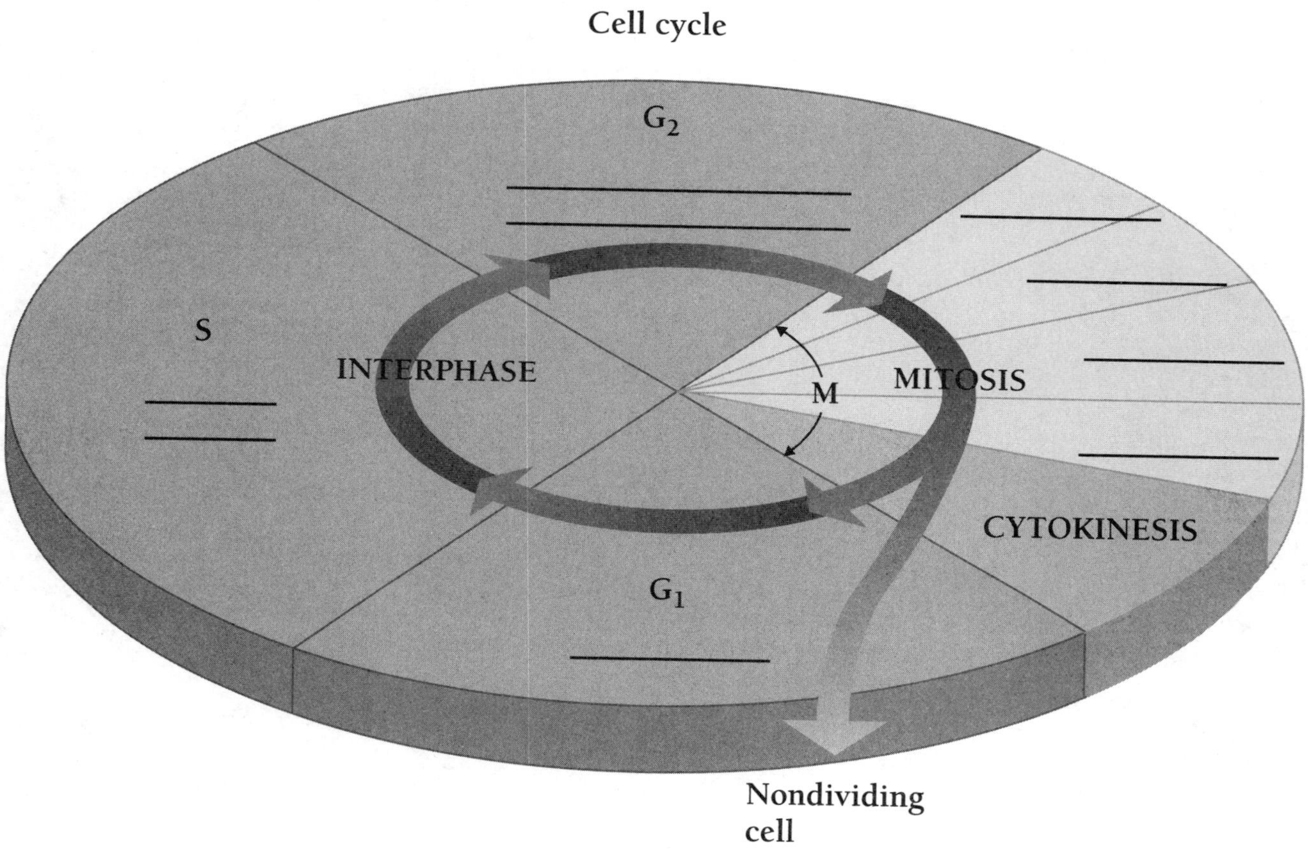

Cell cycle

G_2

S

INTERPHASE

M   MITOSIS

CYTOKINESIS

G_1

Nondividing
cell

Figure 8.8   Mitosis

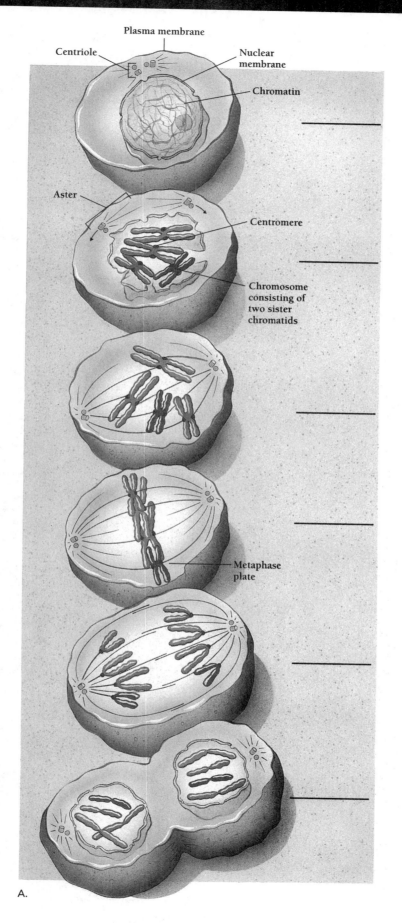

Plasma membrane

Centriole

Nuclear membrane

Chromatin

Aster

Centromere

Chromosome consisting of two sister chromatids

Metaphase plate

A.

Figure 8.13a   Plant Wall Formation Between Daughter Cells

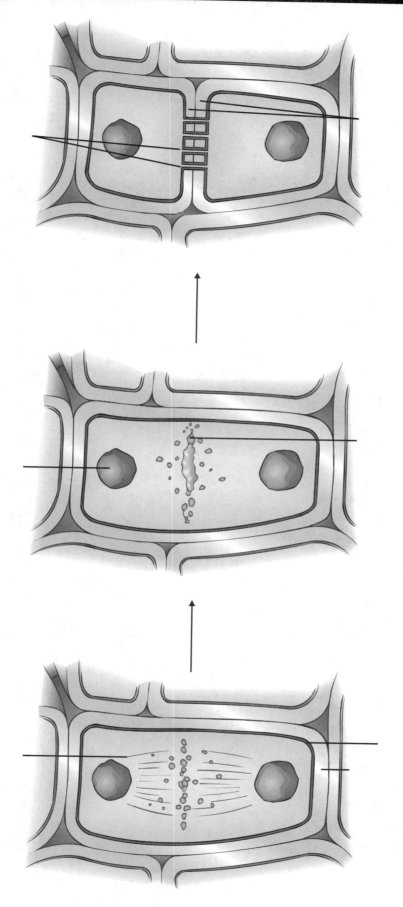

A.

Figure 9.8a   Meiosis I

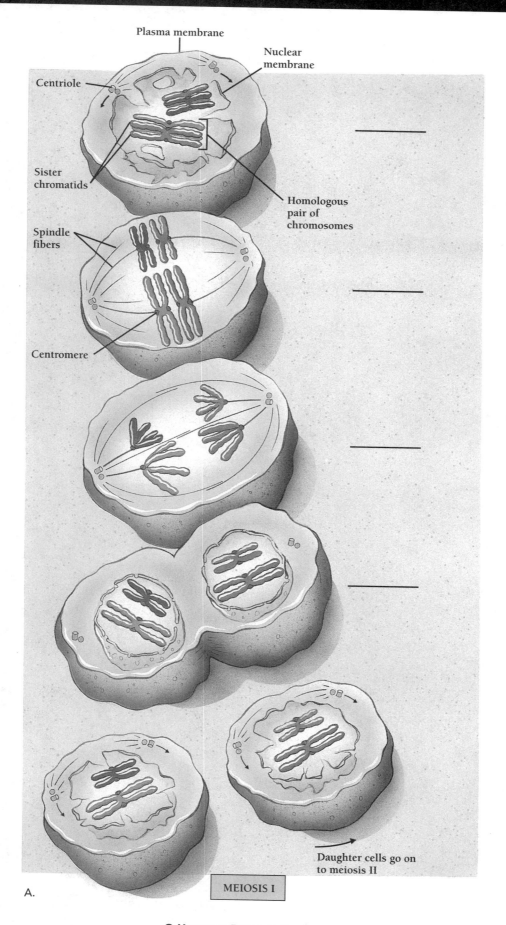

Plasma membrane

Nuclear membrane

Centriole

Sister chromatids

Homologous pair of chromosomes

Spindle fibers

Centromere

Daughter cells go on to meiosis II

A.

MEIOSIS I

Figure 9.8b   Meiosis II

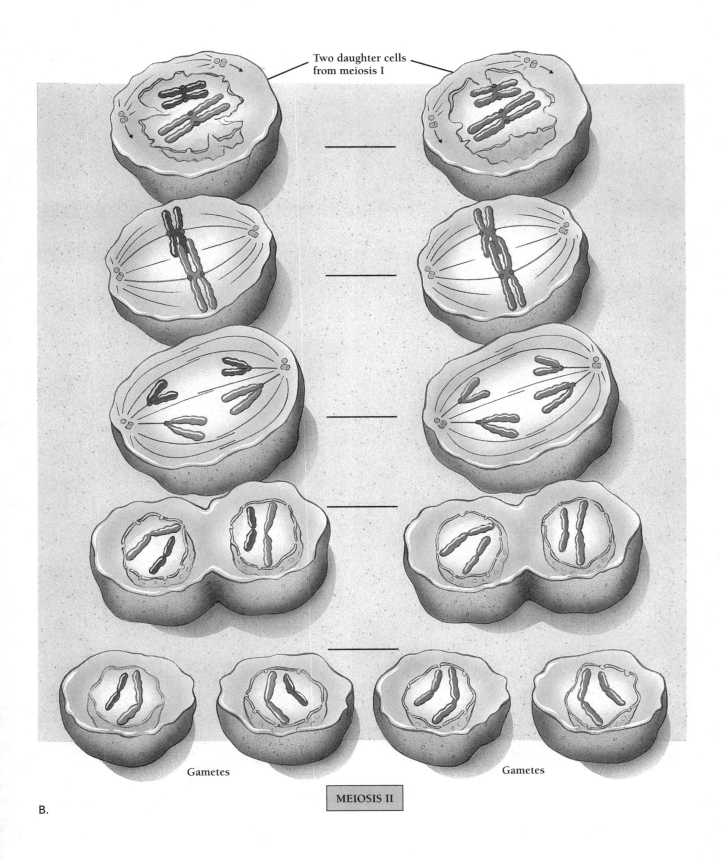

Two daughter cells
from meiosis I

Gametes   Gametes

MEIOSIS II

B.

Figure 11.10   Ribosomes

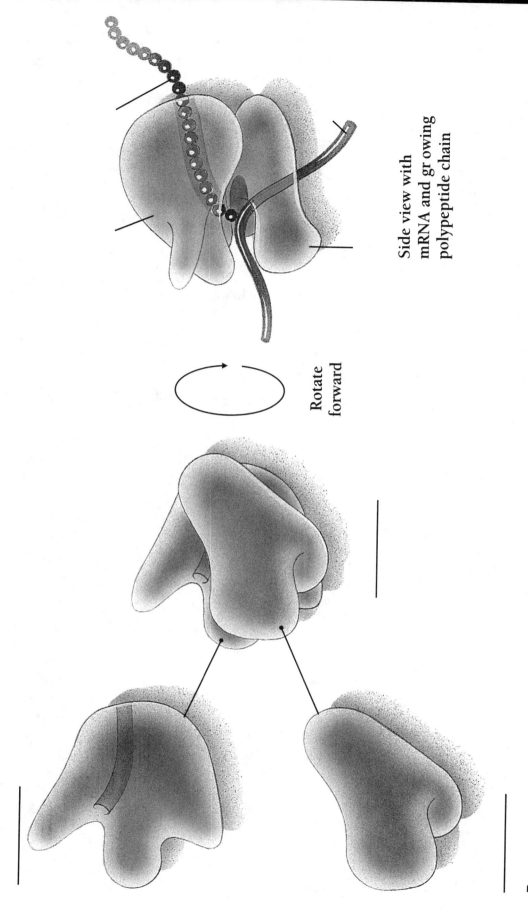

Side view with mRNA and growing polypeptide chain

Rotate forward

B.

Figure 15.2   The Tree of Life

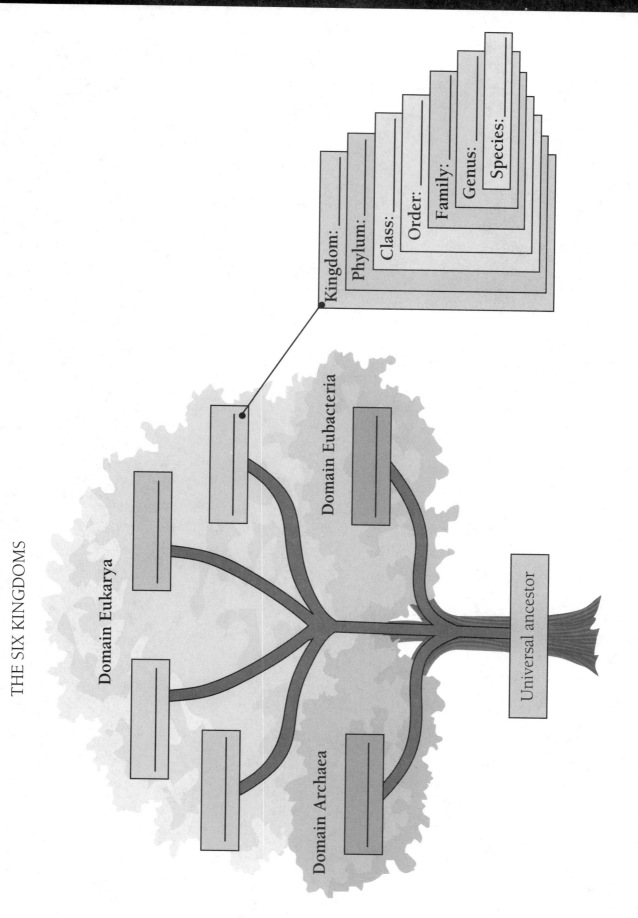

THE SIX KINGDOMS

Kingdom: _____
Phylum: _____
Class: _____
Order: _____
Family: _____
Genus: _____
Species: _____

Domain Eukarya

Domain Eubacteria

Domain Archaea

Universal ancestor

Figure 17.2 The Diversity of Life

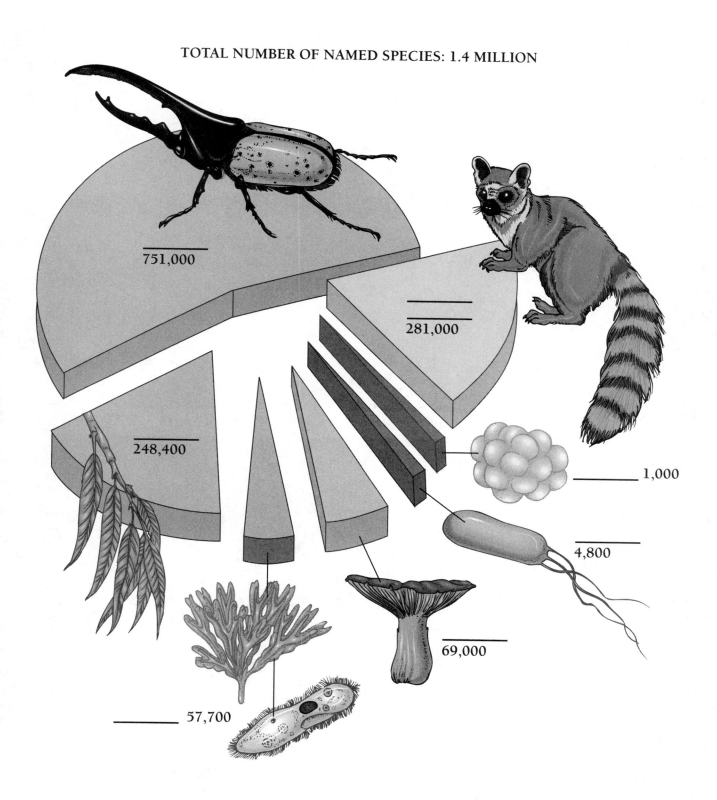

TOTAL NUMBER OF NAMED SPECIES: 1.4 MILLION

751,000

281,000

248,400

1,000

4,800

69,000

57,700

Figure 19.7 A Cladogram for Five Primates

Figure 20.6   Structure of the Bacillus

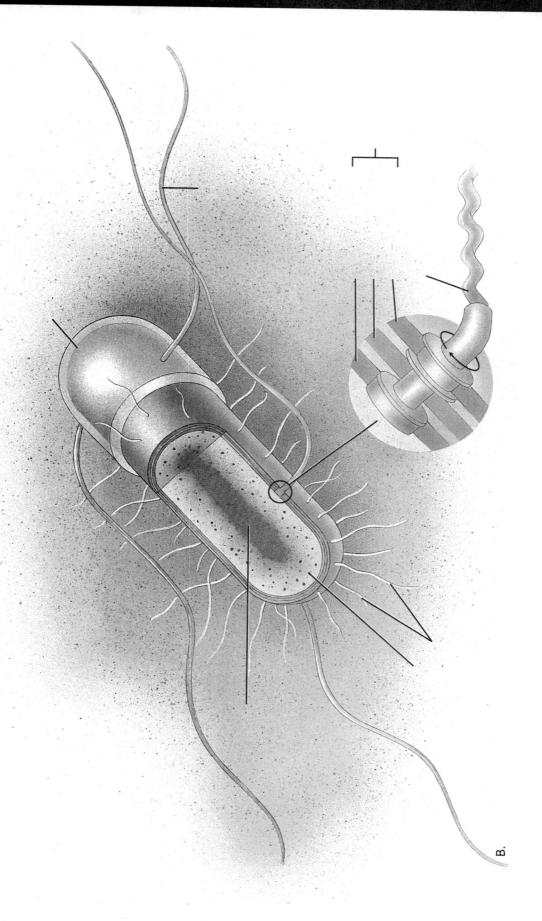

B.

Figure 21.17   The Kingdom of Fungi

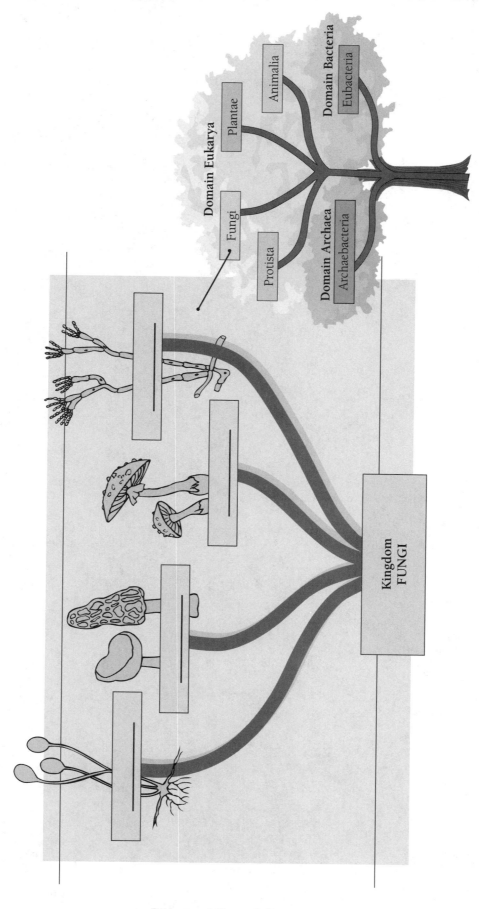

Figure 21.26 Mycorrizae Growth

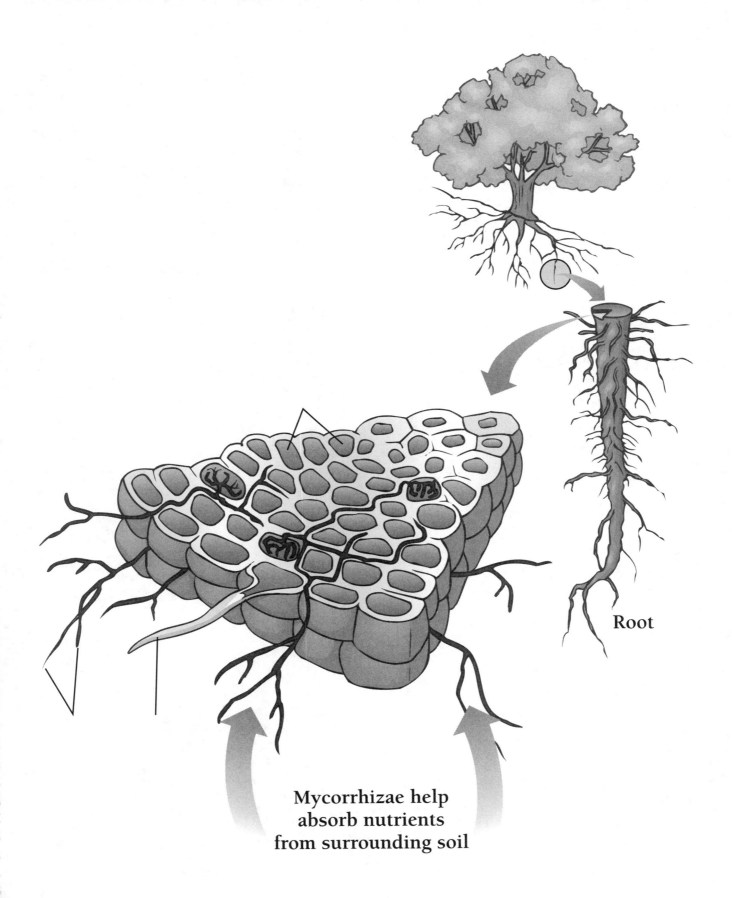

Root

Mycorrhizae help
absorb nutrients
from surrounding soil

Figure 22.1 Linnaeus's Comparison of Flowers and Animals

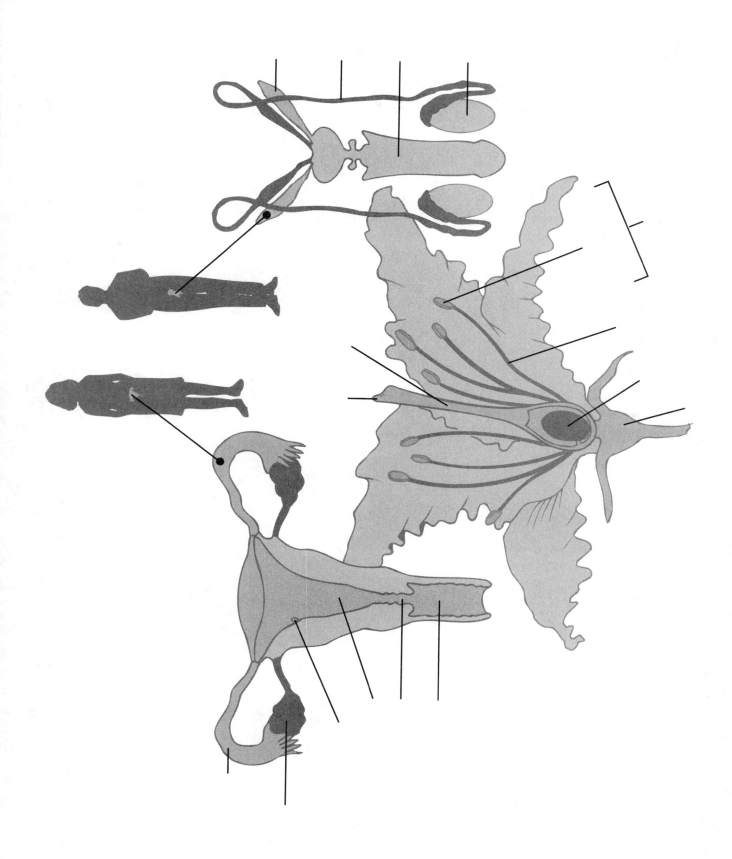

Figure 22.7 Xylem and Phloem

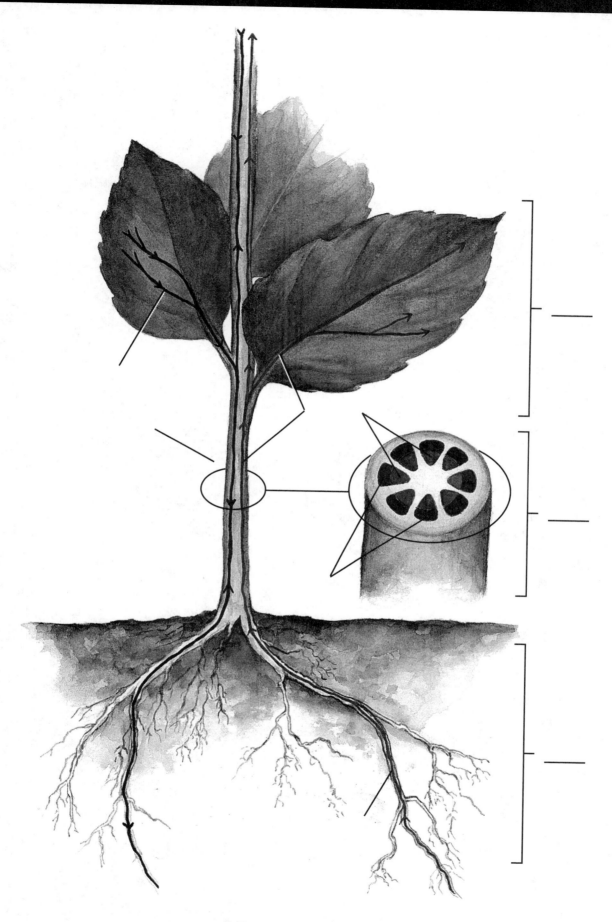

Figure 23.5 Eumetazoan Embryo Development

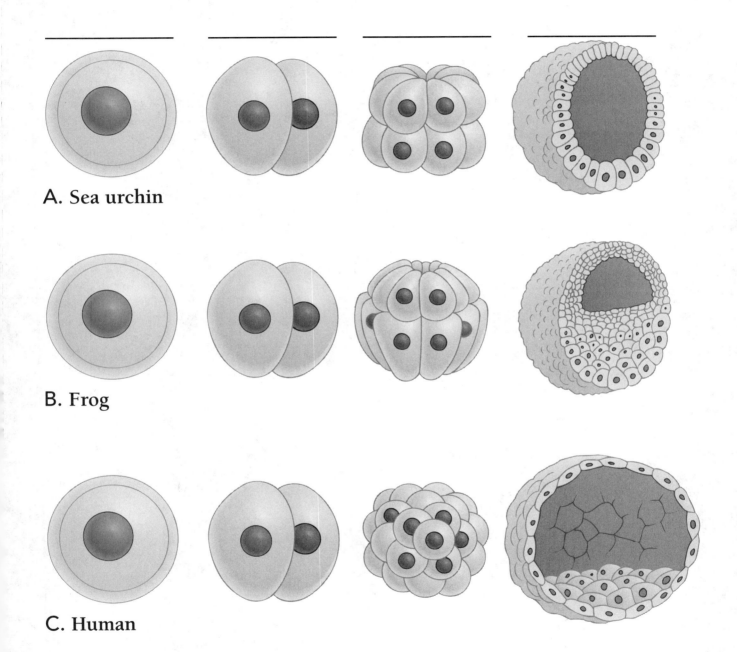

A. Sea urchin

B. Frog

C. Human

Figure 23.10   Cnidarian Body Plans

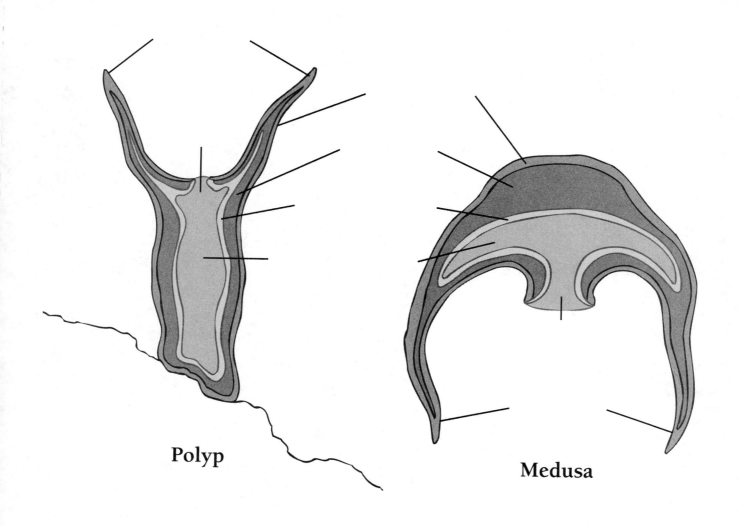

Polyp

Medusa

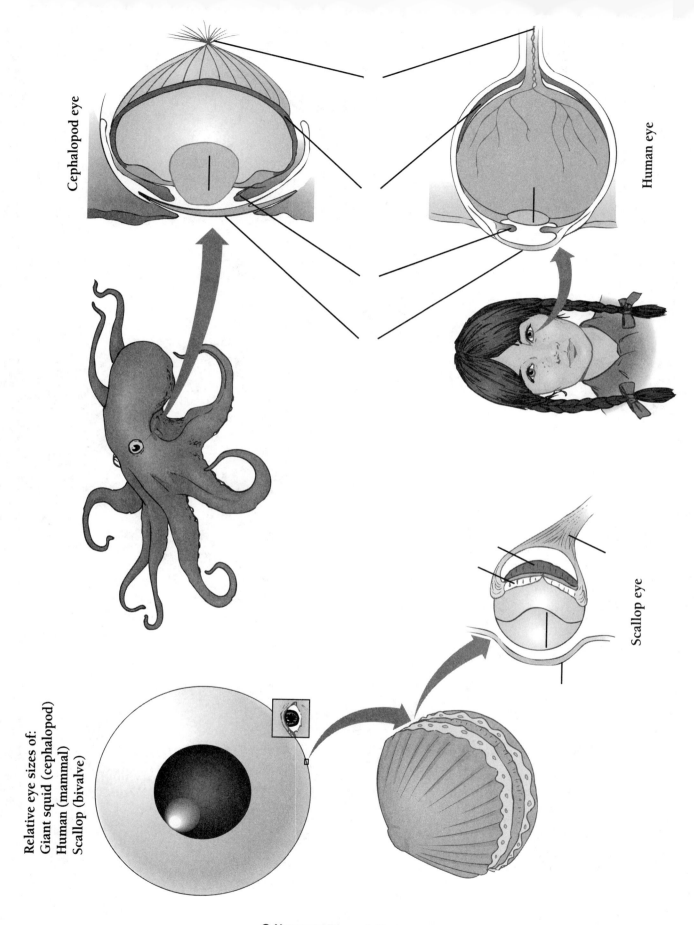

Cephalopod eye

Human eye

Scallop eye

Relative eye sizes of:
Giant squid (cephalopod)
Human (mammal)
Scallop (bivalve)

Figure 23.21 Earthworm Anatomy

Figure 23.22 Complete and Incomplete Metamorphosis

A.

B.

Figure 24.1 Arthropod Body Plan vs. Vertebrate Body Plan

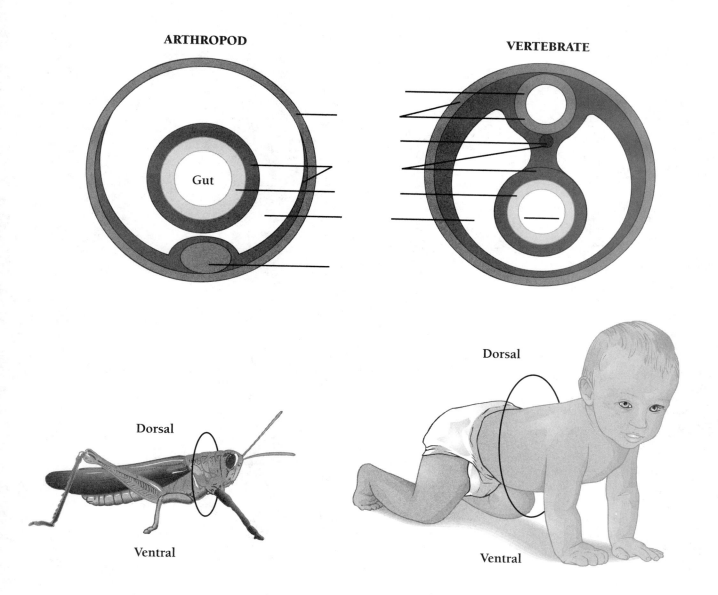

ARTHROPOD

Gut

VERTEBRATE

Dorsal

Ventral

Dorsal

Ventral

Figure 24.4   The Sea Star's Water Vascular System

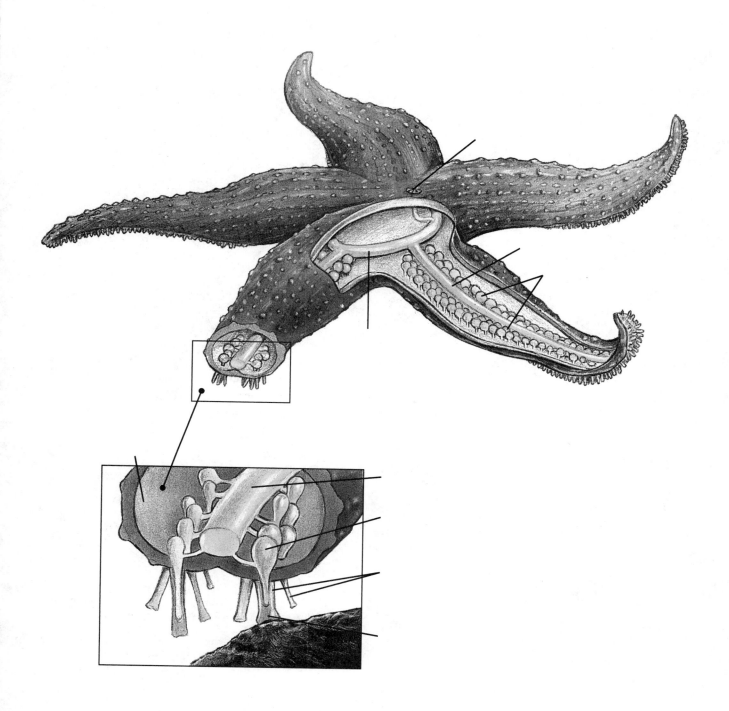

Figure 24.10 Body Plan of Bony Fish

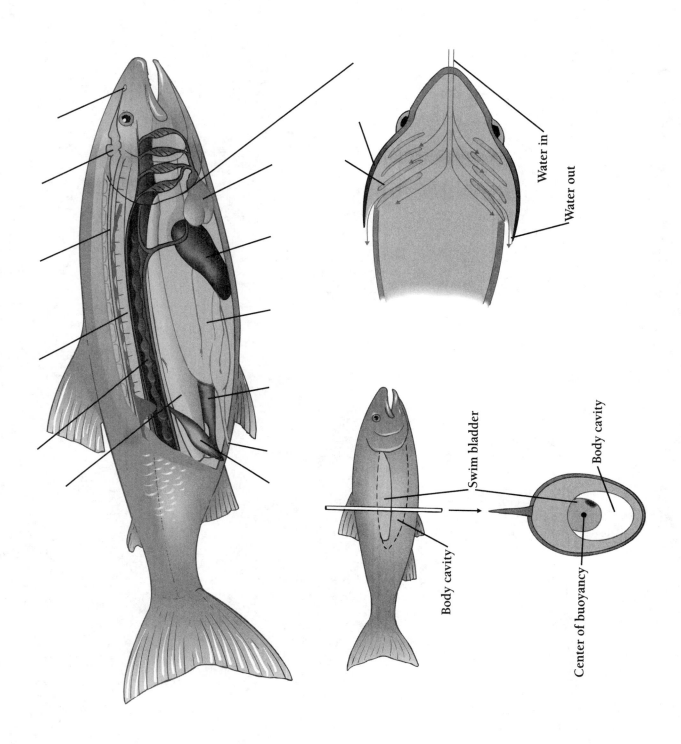

Water in

Water out

Swim bladder

Body cavity

Body cavity

Center of buoyancy

Figure 24.18   The Amniotic Egg

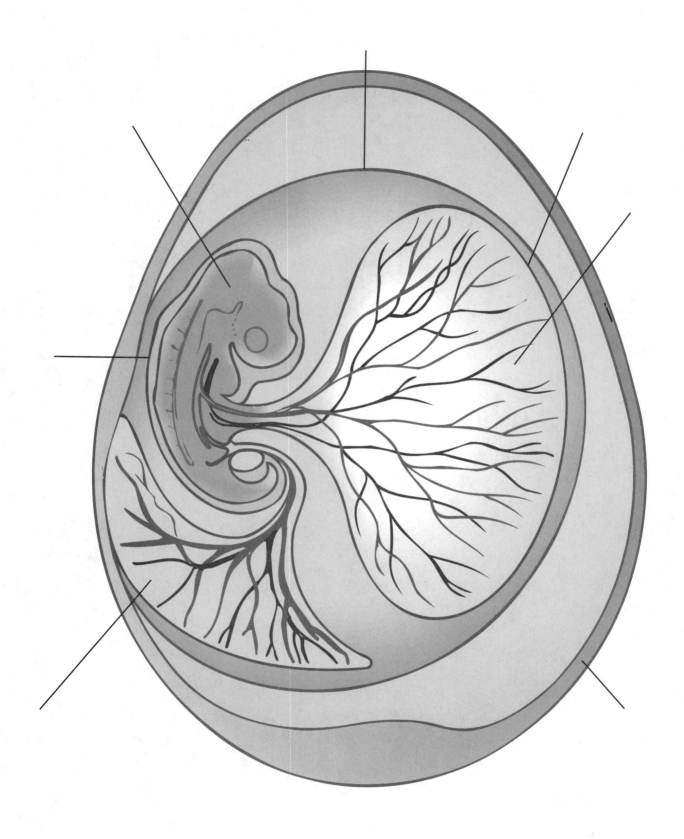

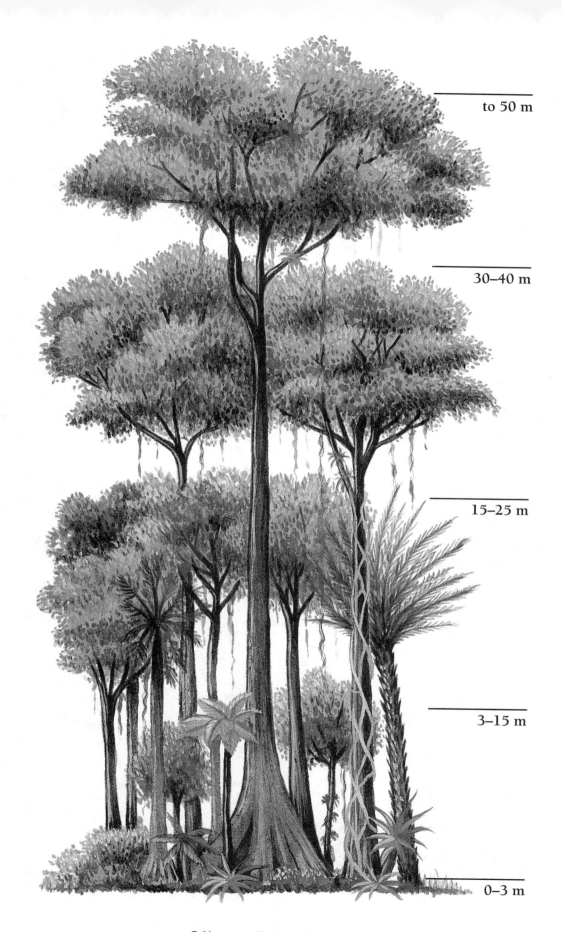

to 50 m

30–40 m

15–25 m

3–15 m

0–3 m

Figure 27.19 Dividing the Ocean into Zones

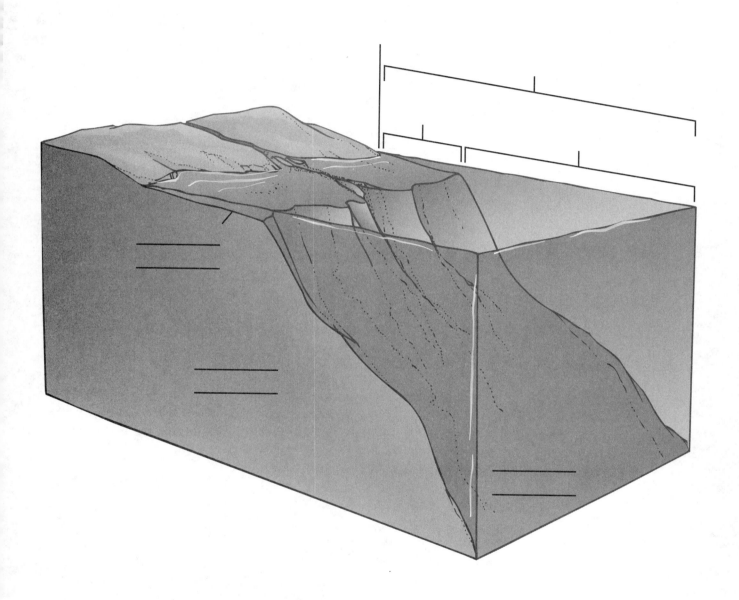

Figure 28.6   Survivorship

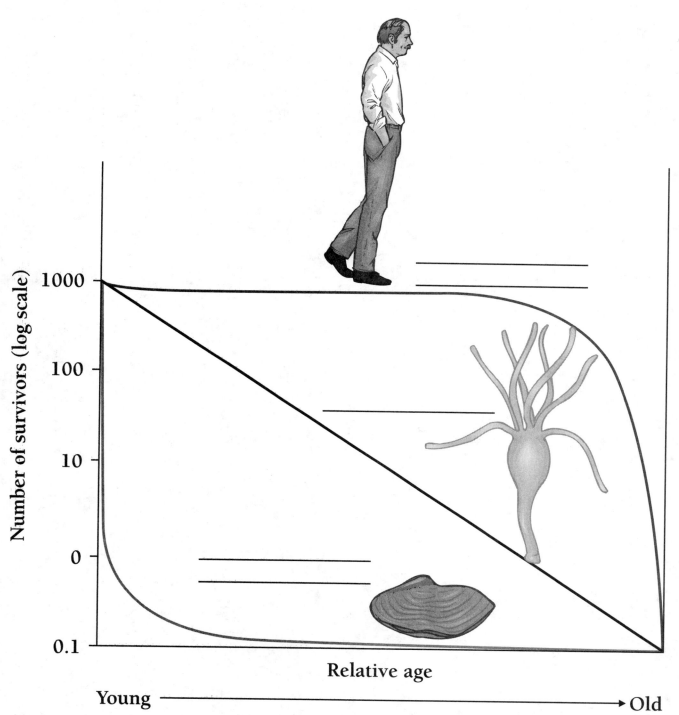

Number of survivors (log scale)

1000

100

10

0

0.1

Relative age

Young ⟶ Old

Survivorship curves: Types I, II, and III

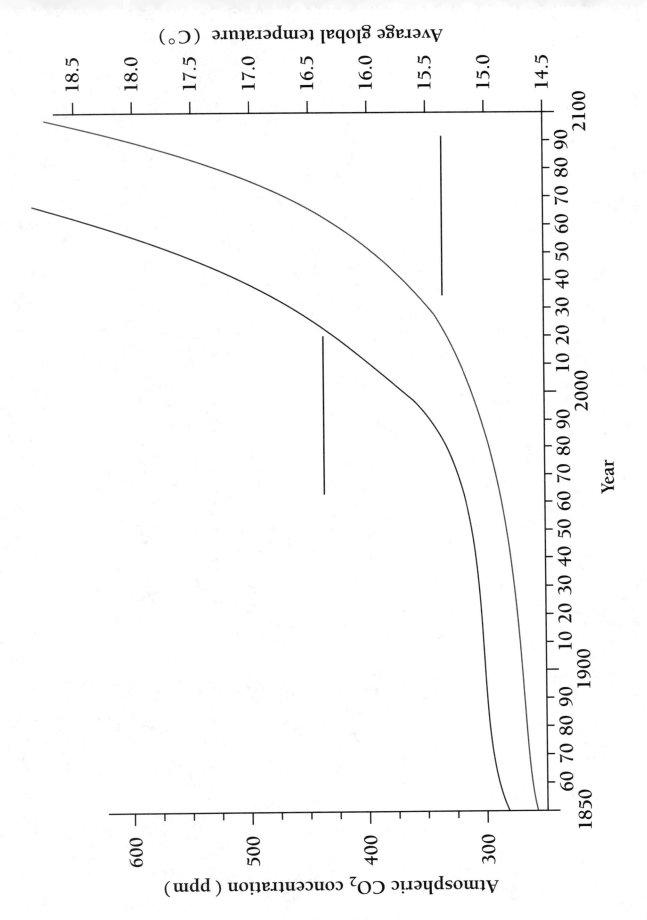

75

Figure 28.14　How Fluorocarbons Breakdown the Ozone Layer

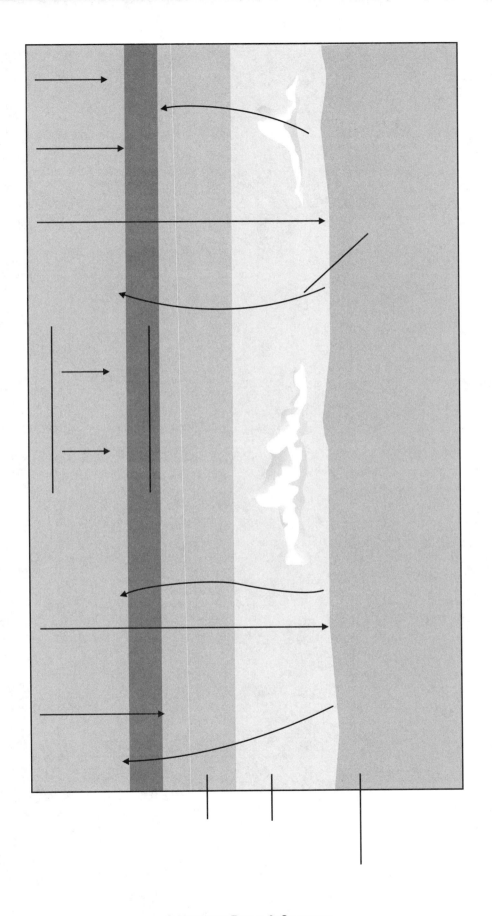

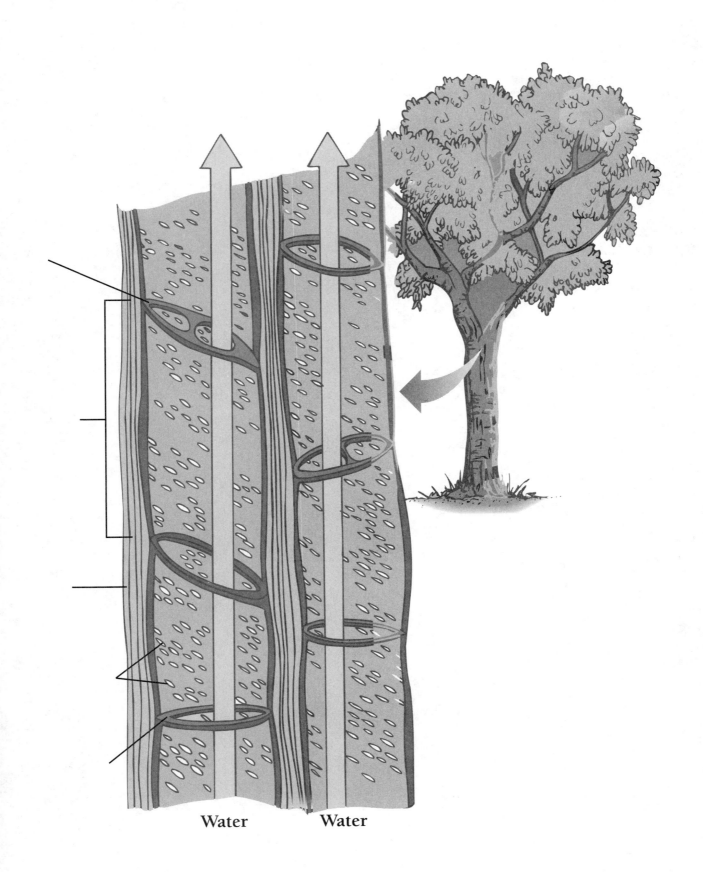

Water   Water

Figure 30.9   The Two Paths of Water Through the Roots

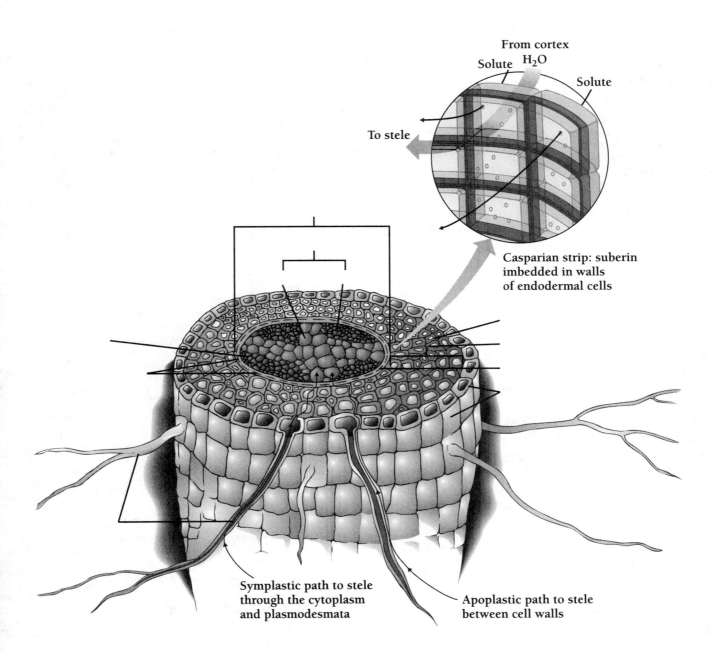

From cortex

H_2O

Solute

Solute

To stele

Casparian strip: suberin imbedded in walls of endodermal cells

Symplastic path to stele through the cytoplasm and plasmodesmata

Apoplastic path to stele between cell walls

Figure 30.12 The Path of Water in the Leaf

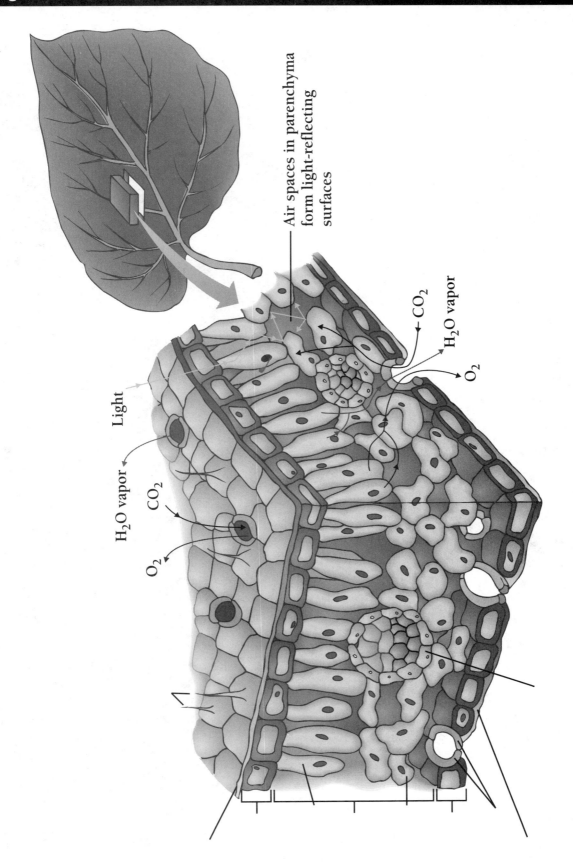

Air spaces in parenchyma form light-reflecting surfaces

CO_2

H_2O vapor

O_2

Light

H_2O vapor

CO_2

O_2

B.

Figure 31.6a   Transport of H₂O and CO₂ through the Stoma

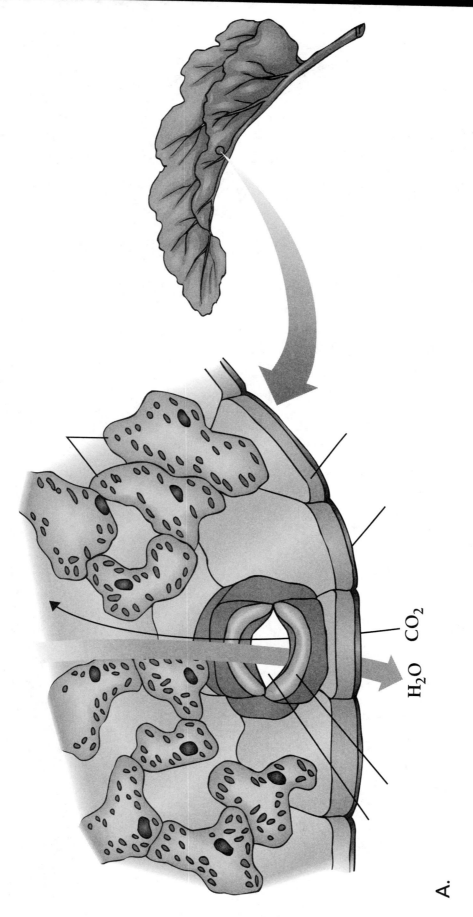

H₂O   CO₂

A.

Figure 32.2 Alternation of Generations in Flowering Plants

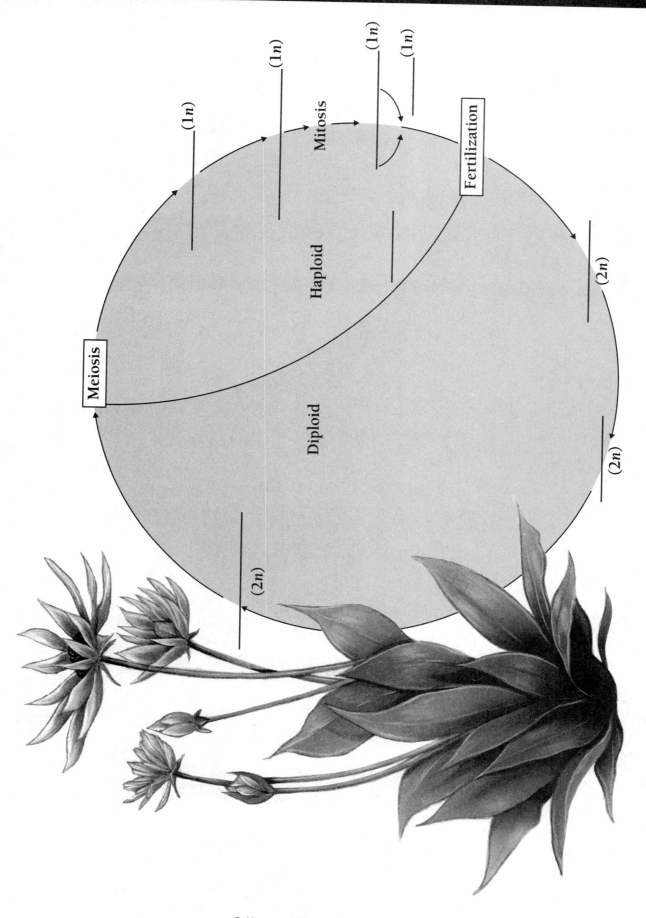

Figure 32.6a   Germination in a Soybean Seed

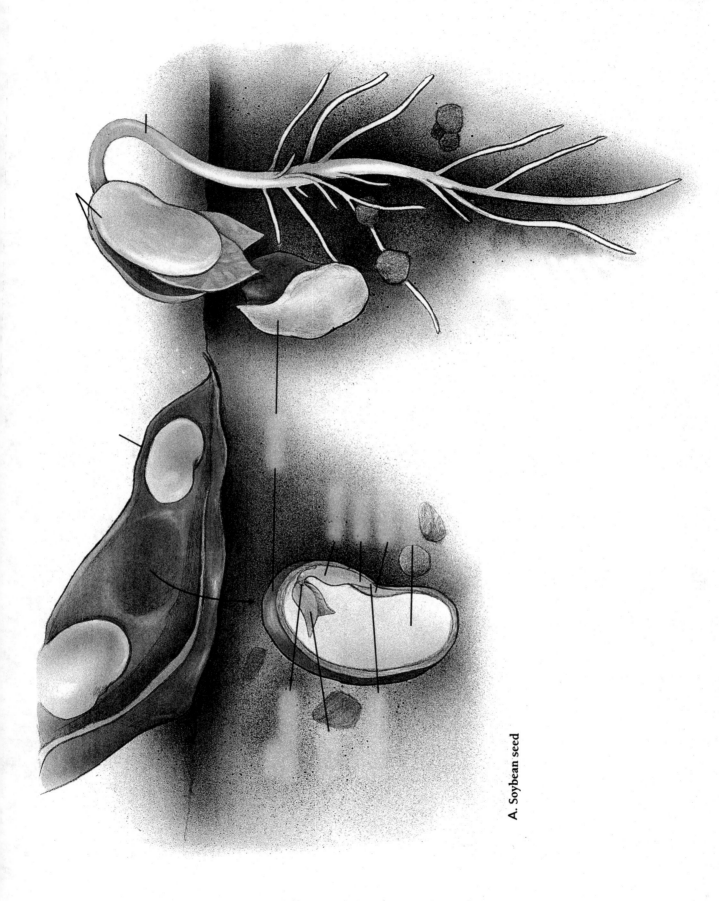

A. Soybean seed

Figure 32.6b Germination in a Corn Seed

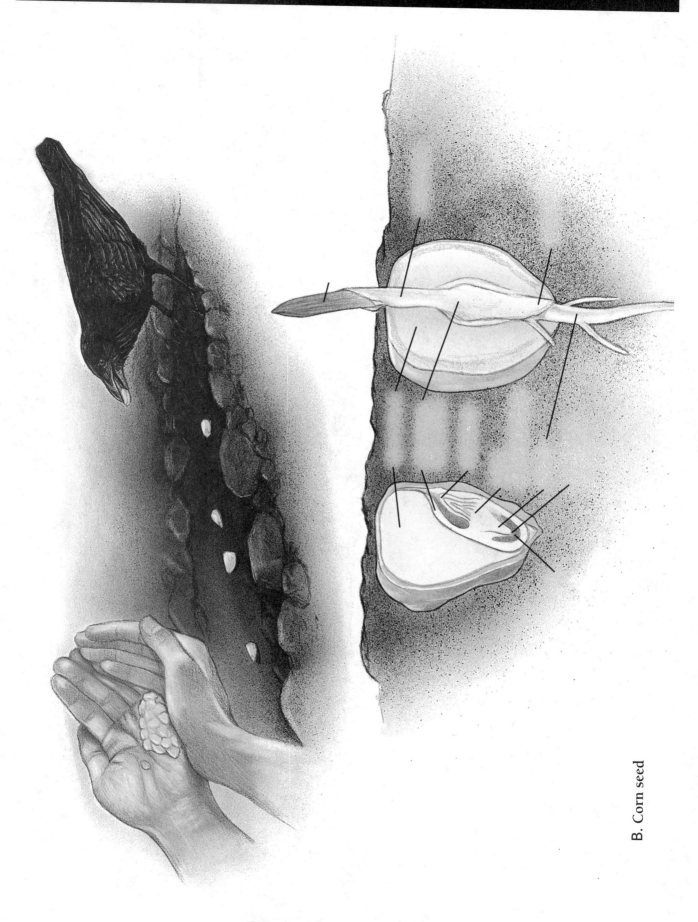

B. Corn seed

Figure 32.10   Growth in a Root

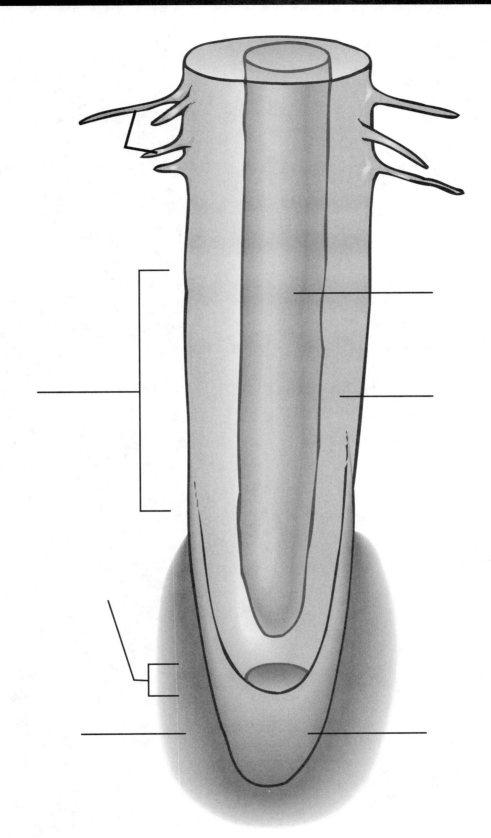

Longitudinal section of a root tip

Figure 32.13 Secondary Growth

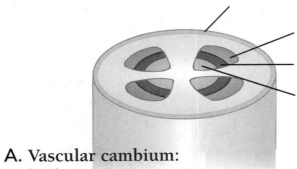

A. Vascular cambium:
herbaceous stem

B. Vascular and cork cambium

Figure 34.7   Skeleton of a Flying Bird

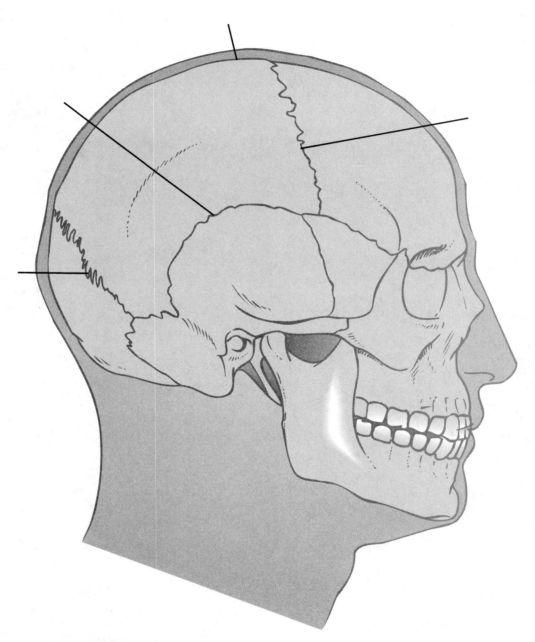

B. Sutures of the skull

ACCESSORY ORGANS:

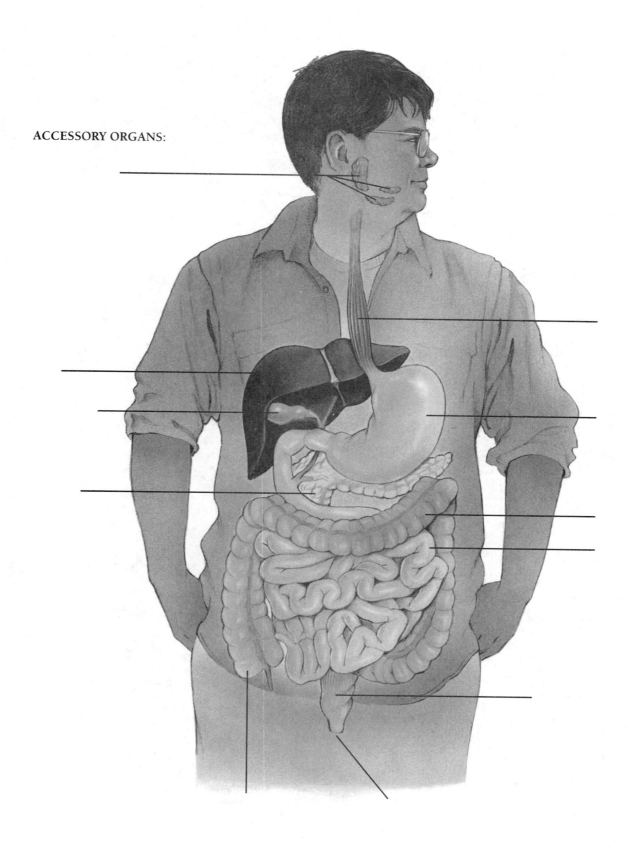

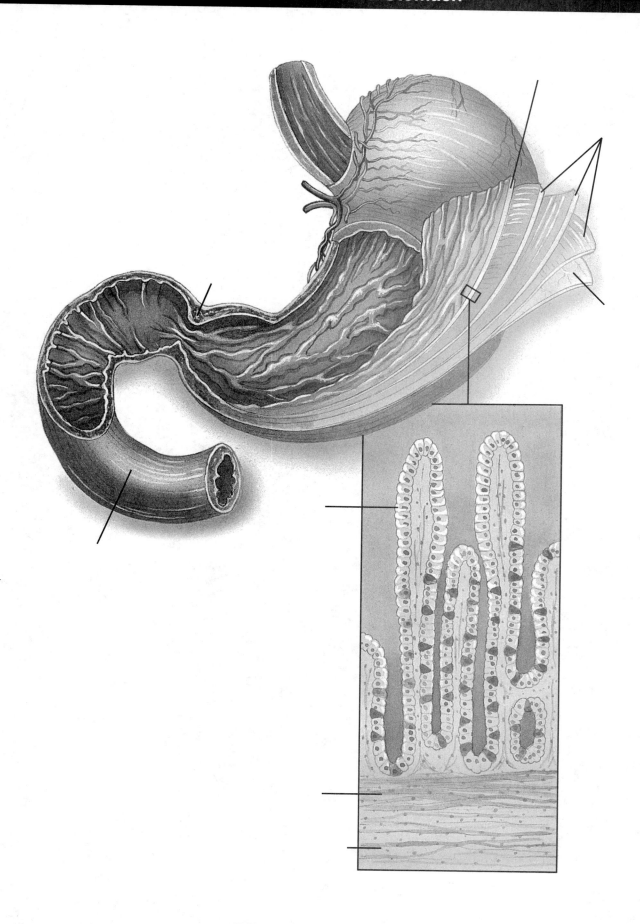

Figure 35.11   The Four Stomachs of a Ruminant

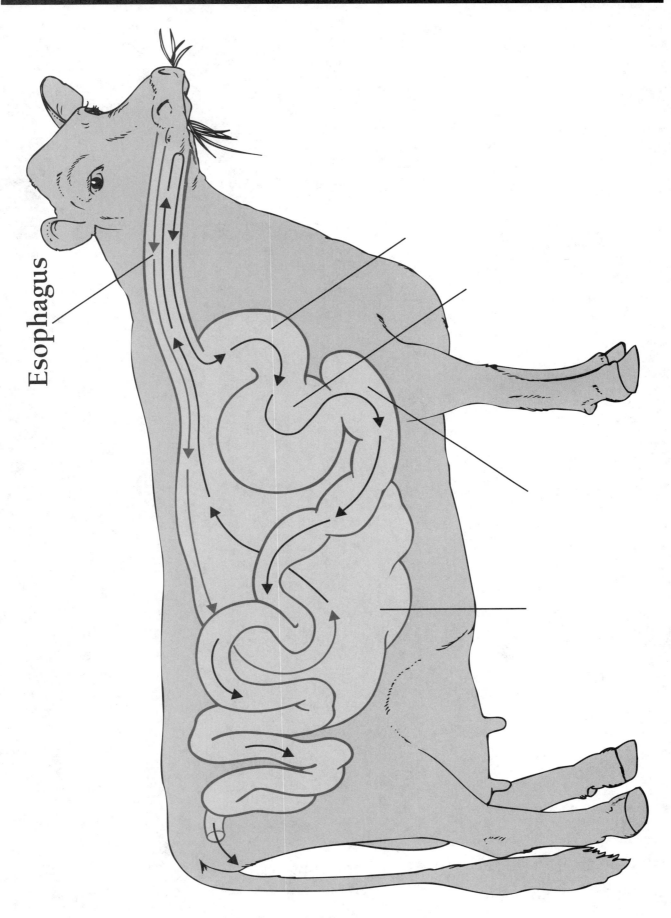

Esophagus

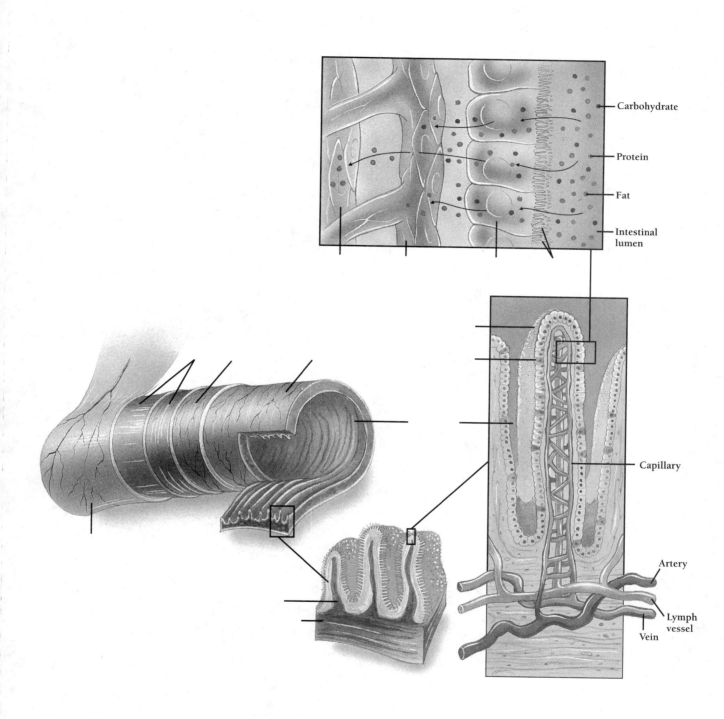

Carbohydrate

Protein

Fat

Intestinal
lumen

Capillary

Artery

Lymph
vessel

Vein

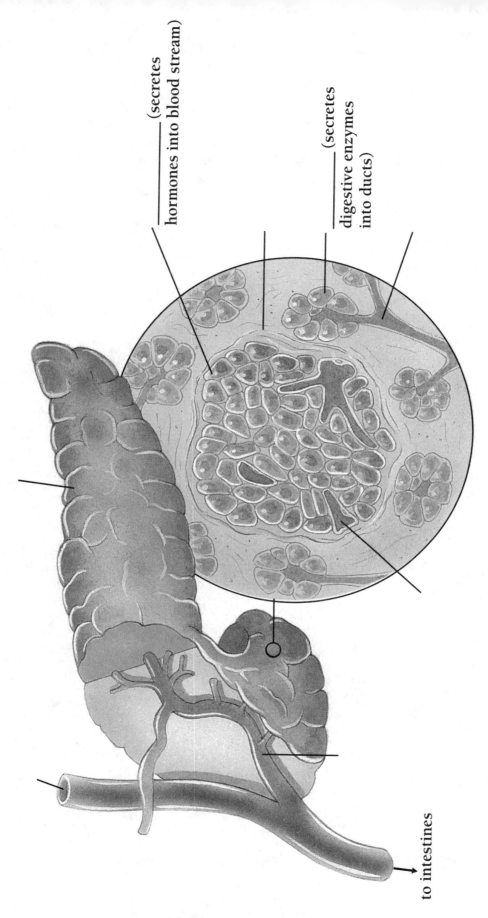

(secretes hormones into blood stream)

(secretes digestive enzymes into ducts)

to intestines

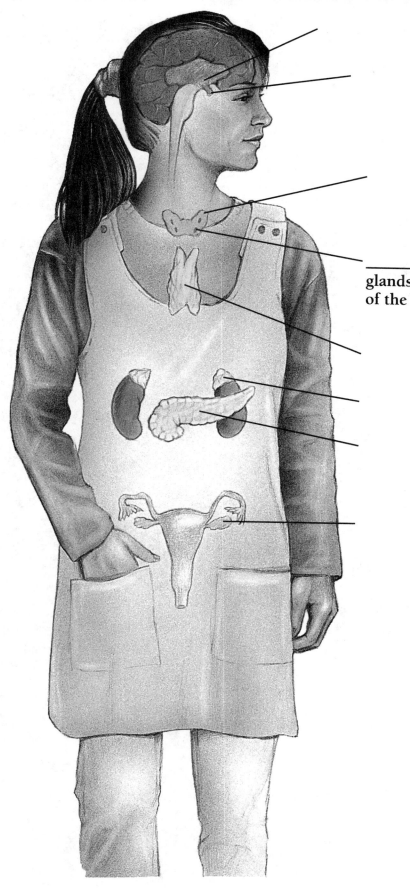

_____ (four small
glands on the dorsal side
of the thyroid gland)

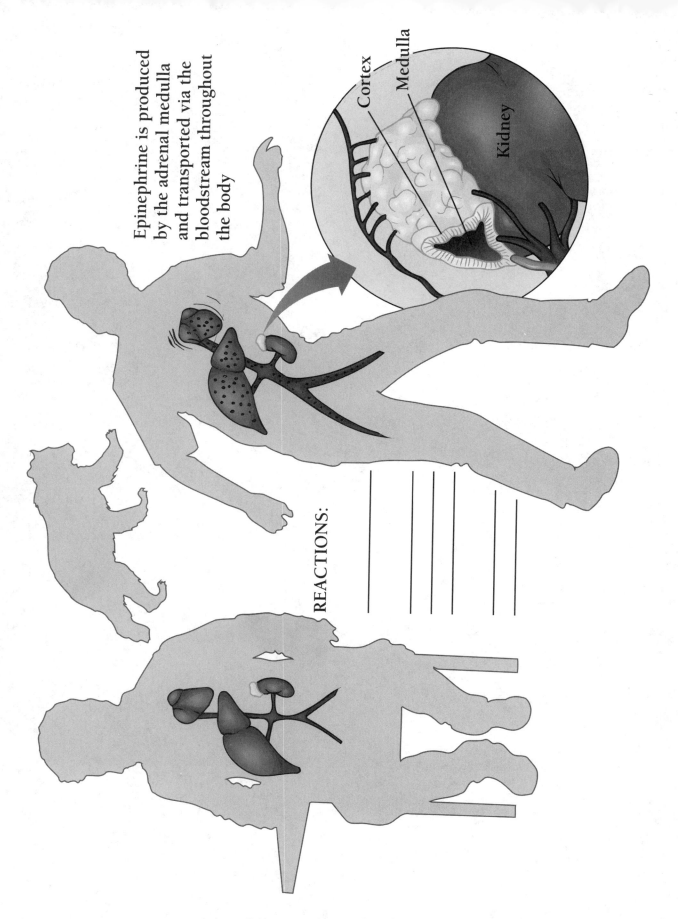

Epinephrine is produced by the adrenal medulla and transported via the bloodstream throughout the body

Cortex

Medulla

Kidney

REACTIONS:

Figure 36.7   The Hypothalamus and Pituitary Glands

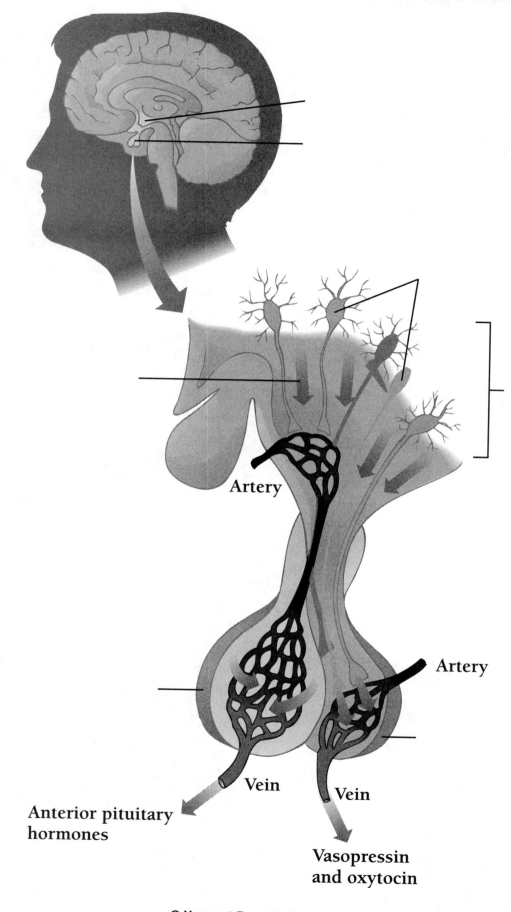

Artery

Artery

Vein

Vein

Anterior pituitary
hormones

Vasopressin
and oxytocin

Figure 37.5   The Human Circulatory System

VEINS

ARTERIES

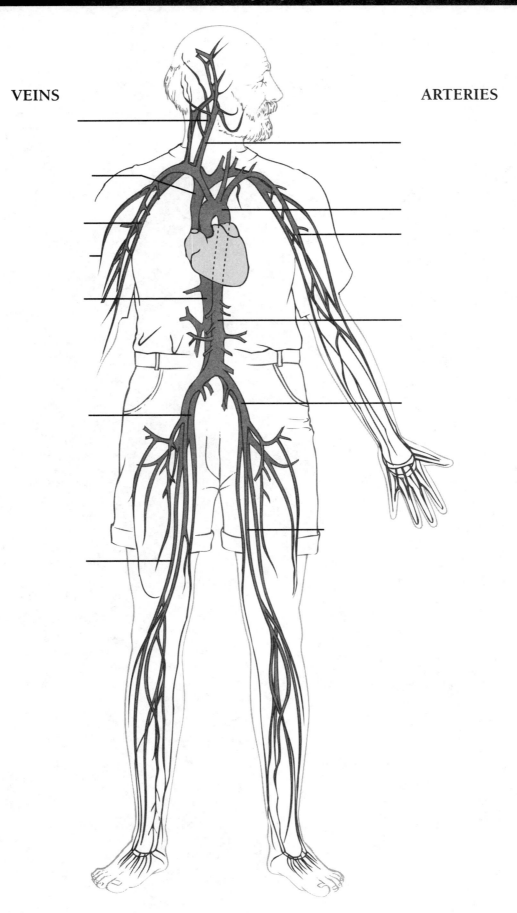

Figure 37.6   The Heart

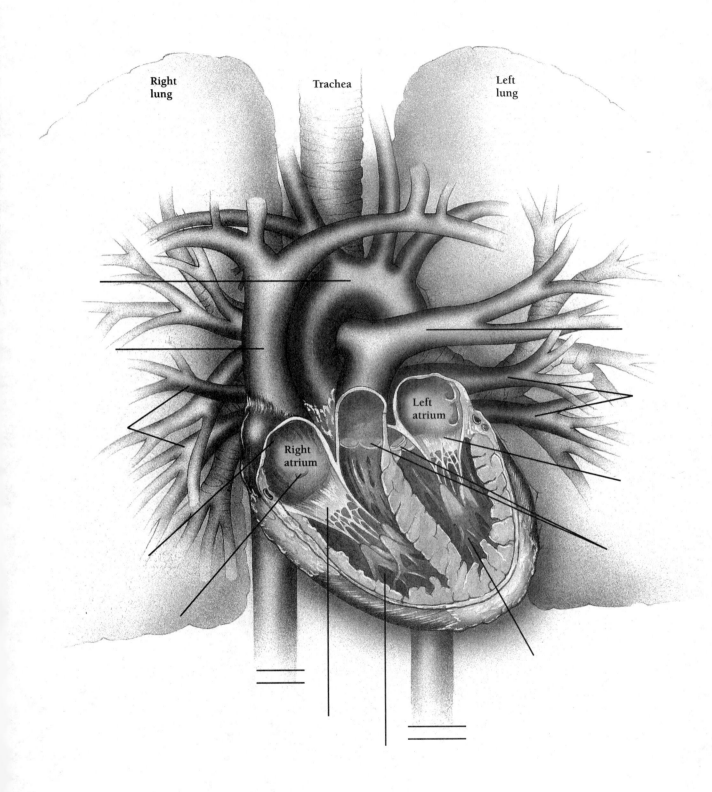

Right
lung

Trachea

Left
lung

Left
atrium

Right
atrium

Figure 37.10   The Lymphatic System

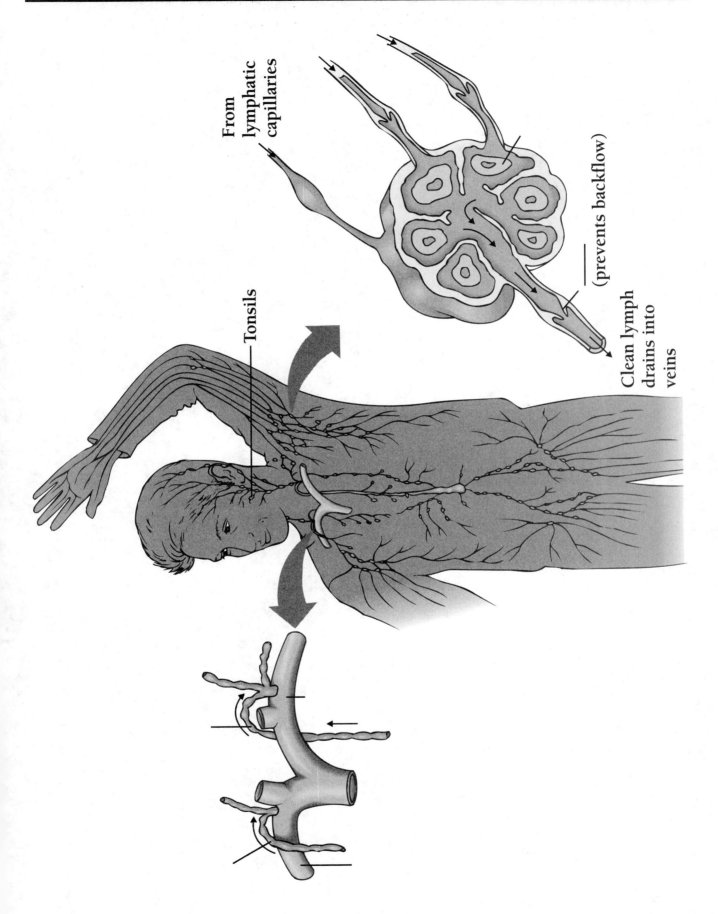

From lymphatic capillaries

(prevents backflow)

Clean lymph drains into veins

Tonsils

Figure 37.11a Blood Vessel Layers

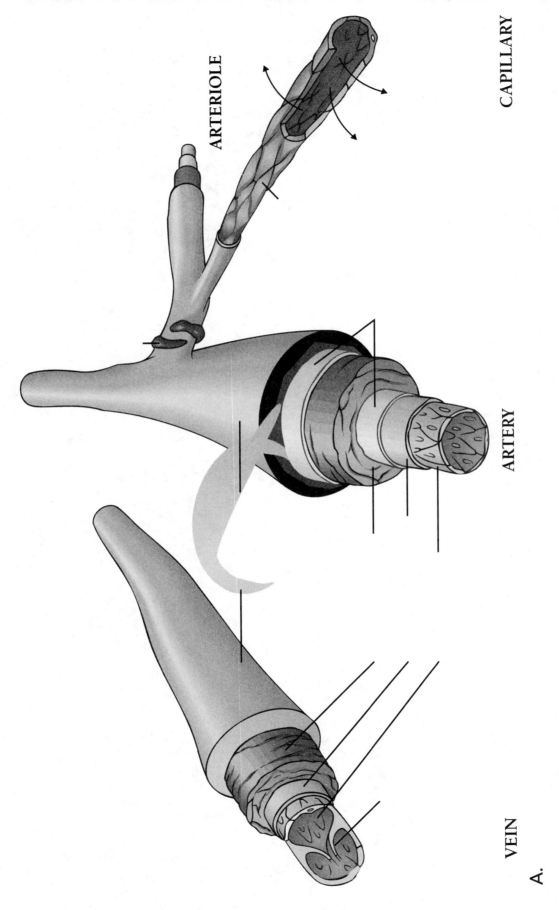

ARTERIOLE

CAPILLARY

ARTERY

VEIN

A.

Death rate per 100,000 population

Year

B.

Figure 38.7   The Human Respiratory System

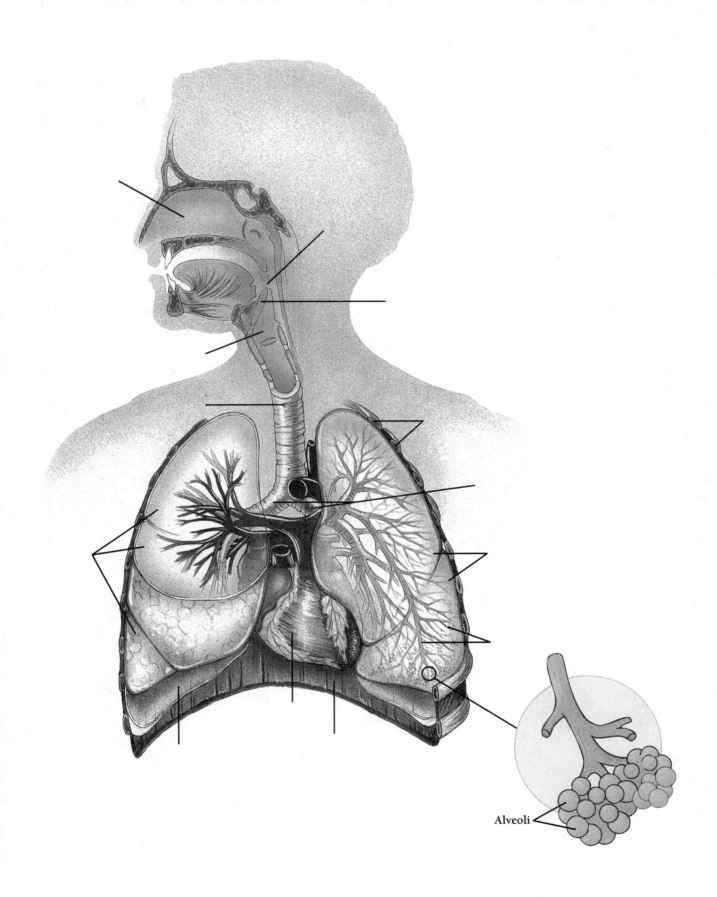

Alveoli

Figure 38.8 Epithelia in the Respiratory Tract

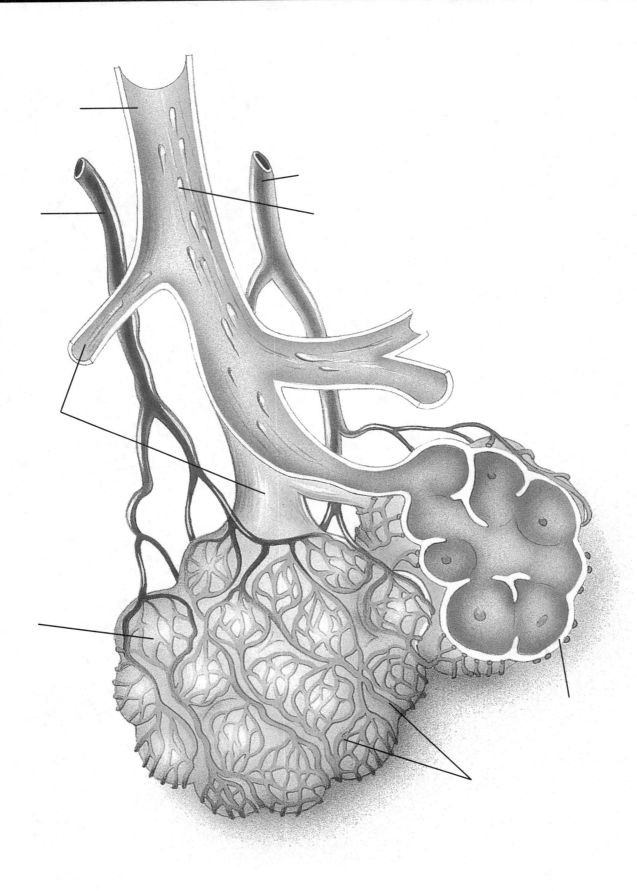

Figure 38.9 Cellular Organization of Alveoli and Blood Vessels

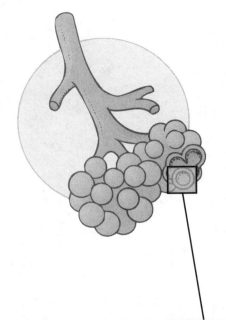

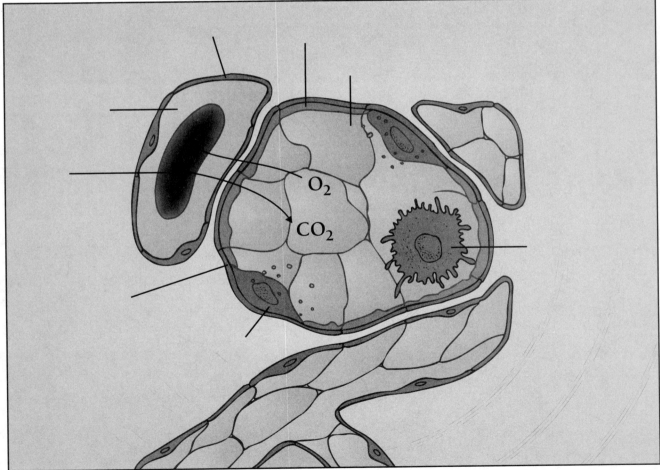

O_2

CO_2

Figure 39.6   The Human Excretory System

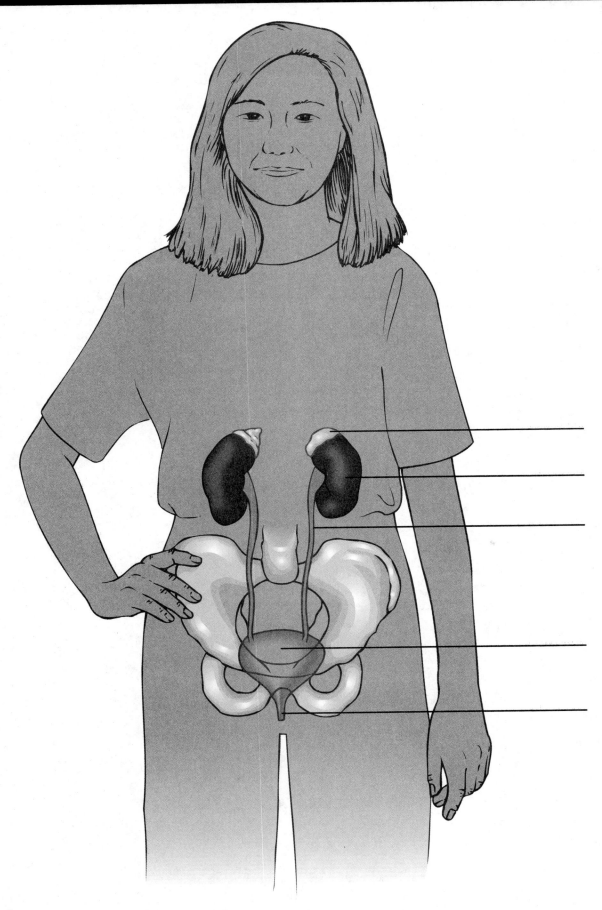

Figure 39.7 (top) Cross Section through a Kidney

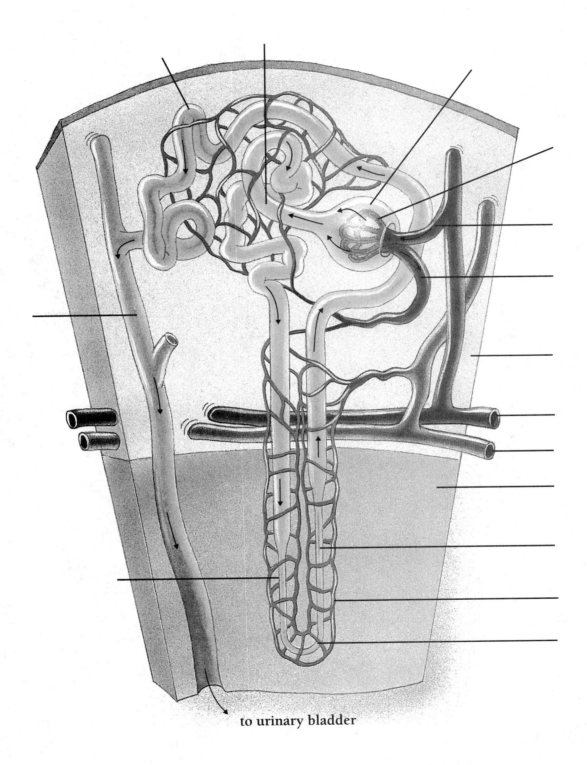

to urinary bladder

Figure 39.7 (bottom)   Cross Section through a Kidney

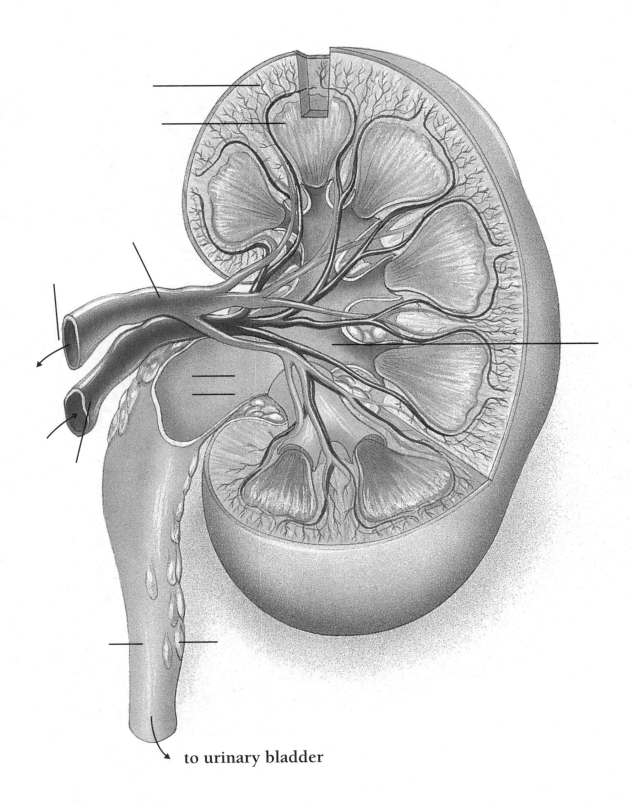

to urinary bladder

Figure 39.8   A Glomerulus

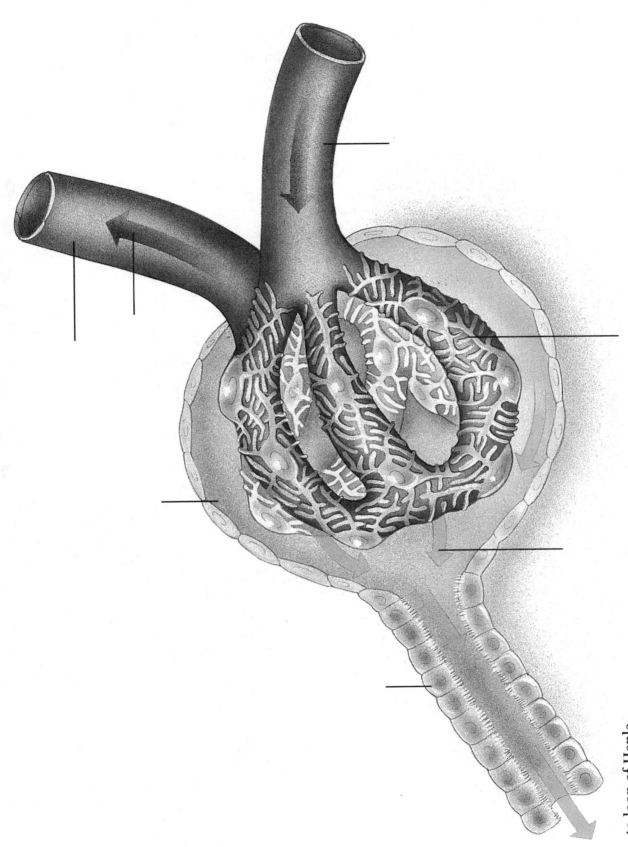

to loop of Henle, distal tubule, and points beyond

Figure 40.3   The Epidermis

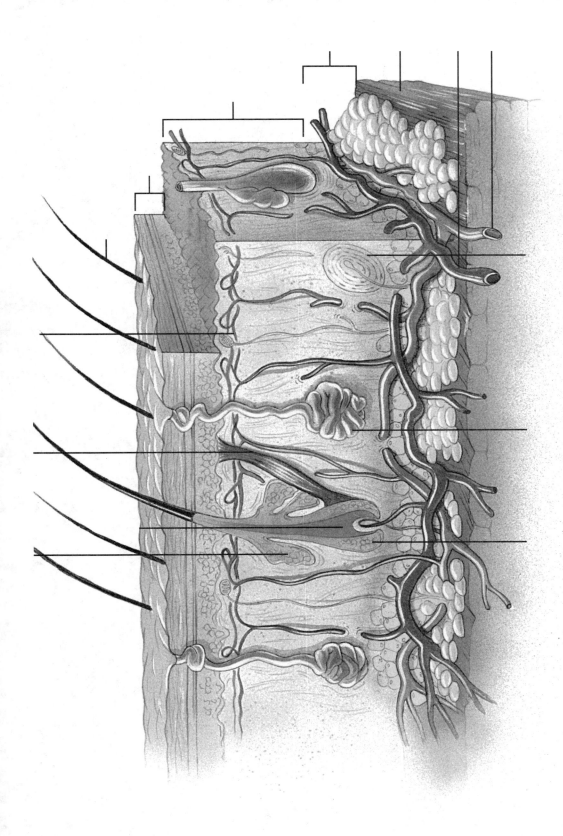

Figure 40.4   Hemostasis

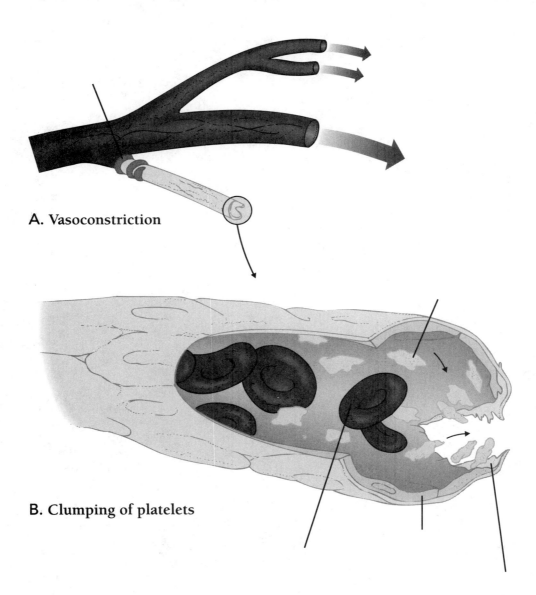

A. Vasoconstriction

B. Clumping of platelets

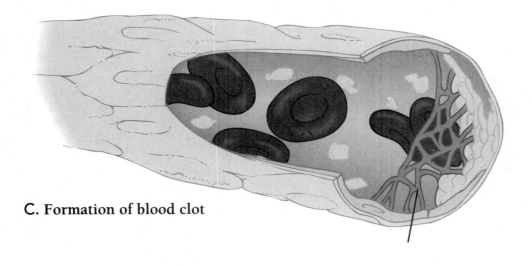

C. Formation of blood clot

Figure 40.5 Tissue Damage

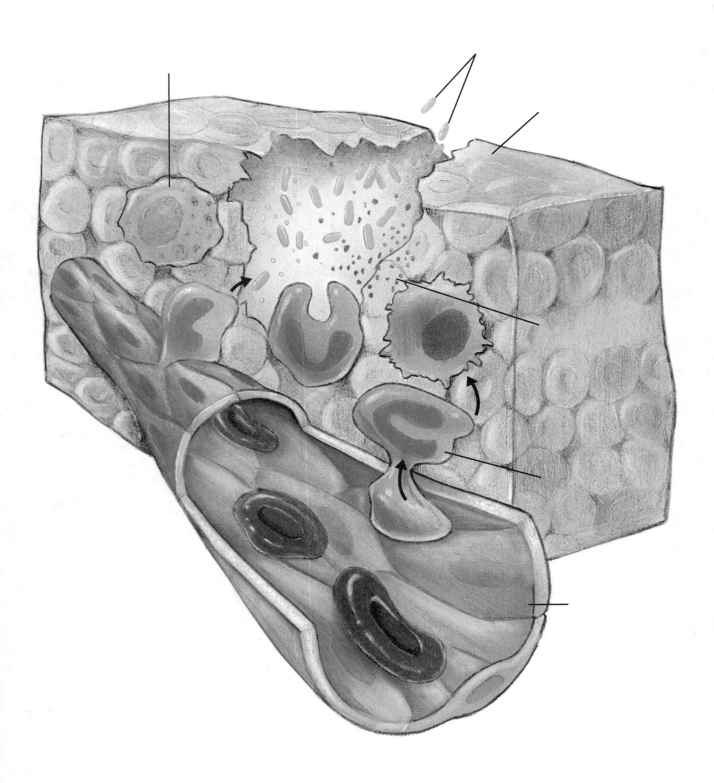

Figure 40.8 Structure of an Antibody Molecule

Figure 41.4   Nerve Cell

B.

Figure 41.8 Synapses

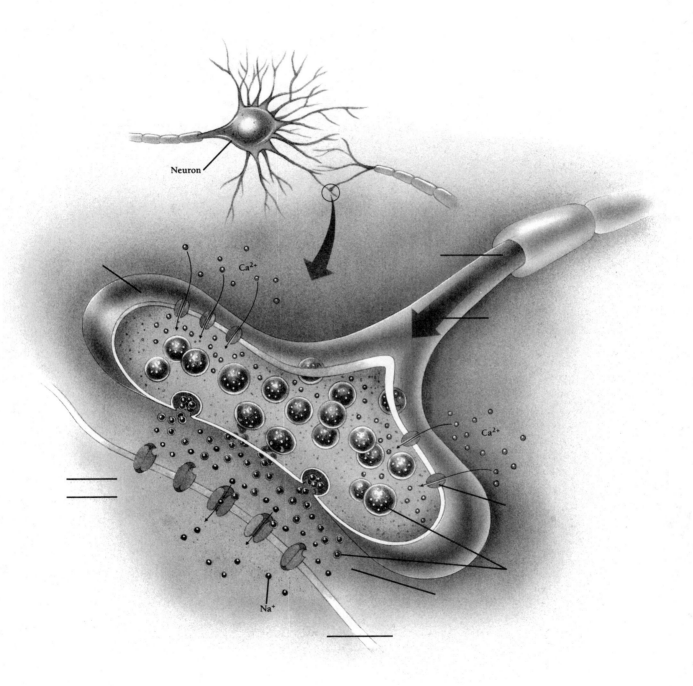

Neuron

Ca²⁺

Ca²⁺

Na⁺

Figure 42.4   Nervous System of a Human

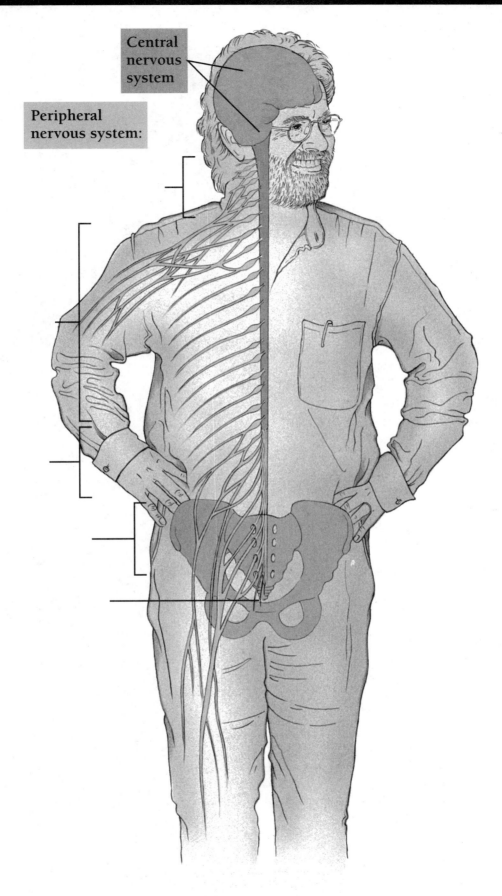

Central
nervous
system

Peripheral
nervous system:

Figure 42.6 Mammalian Nervous System

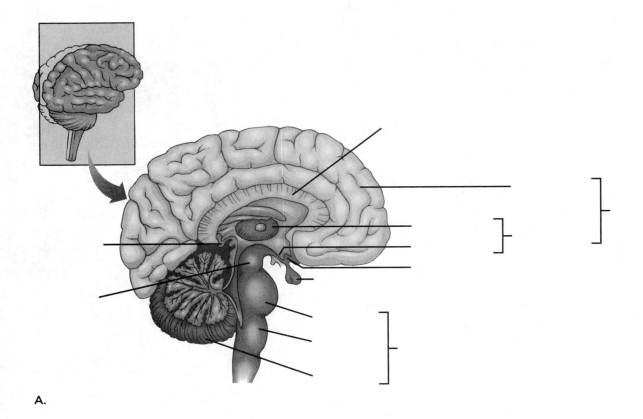

A.

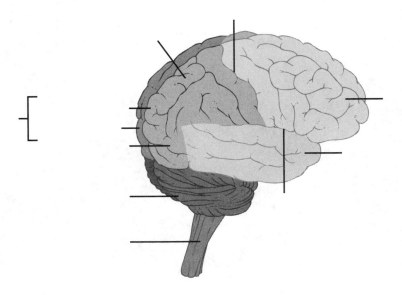

Figure 42.8   The Human Ear

External ear   Middle ear   Inner ear

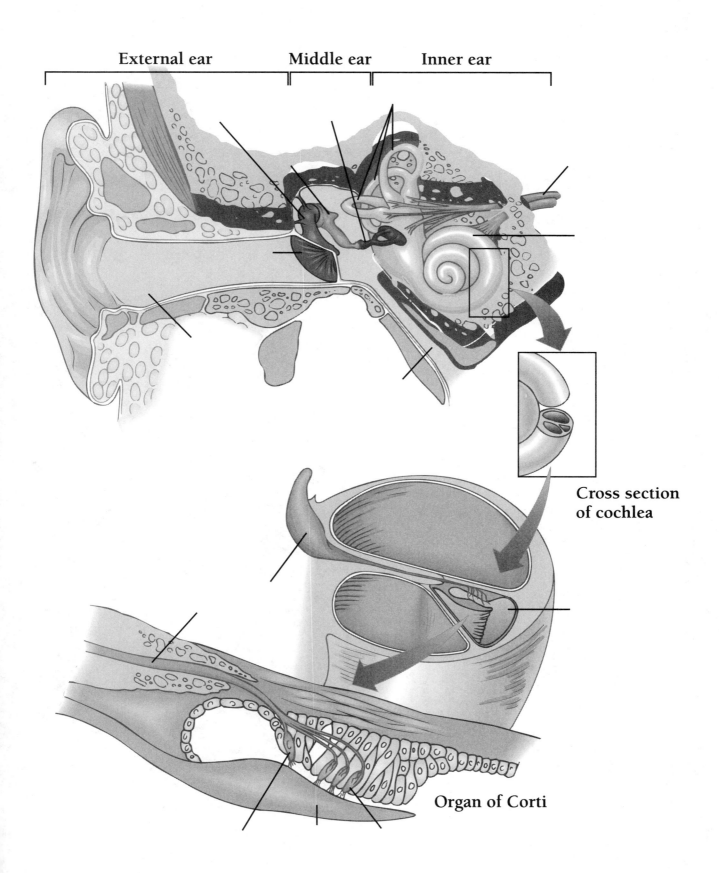

Cross section
of cochlea

Organ of Corti

Figure 42.10   The Human Eye

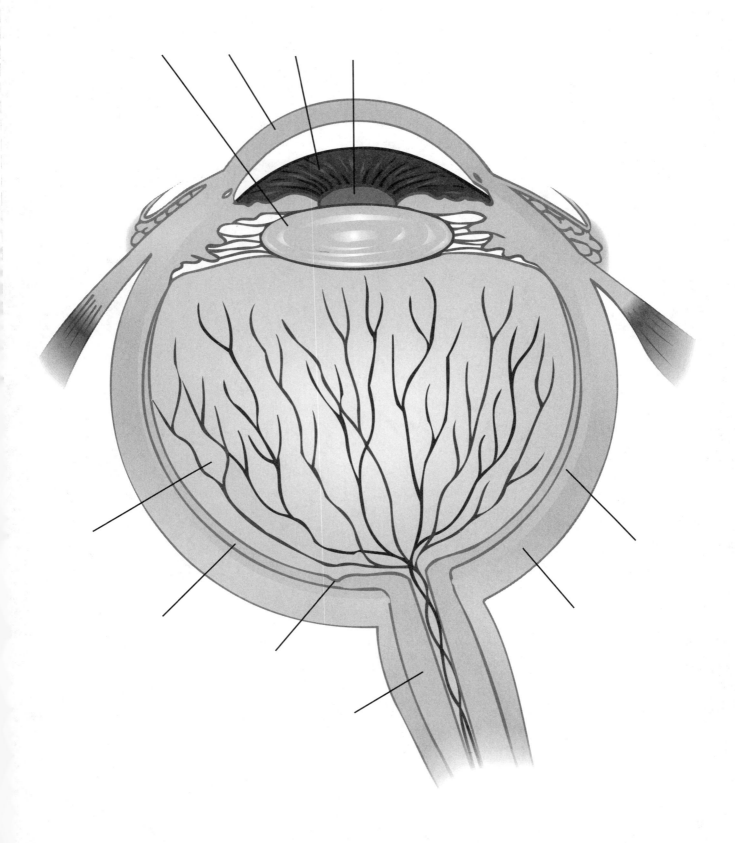

Figure 42.12 Rods and Cones

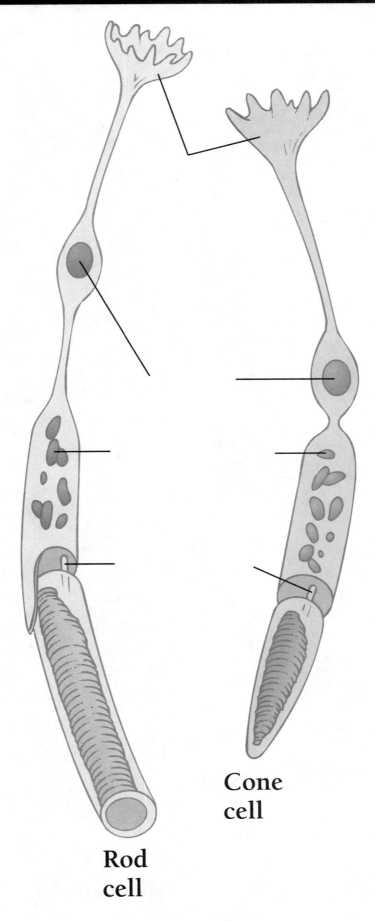

Rod
cell

Cone
cell

Figure 42.13 Organization of the Retina

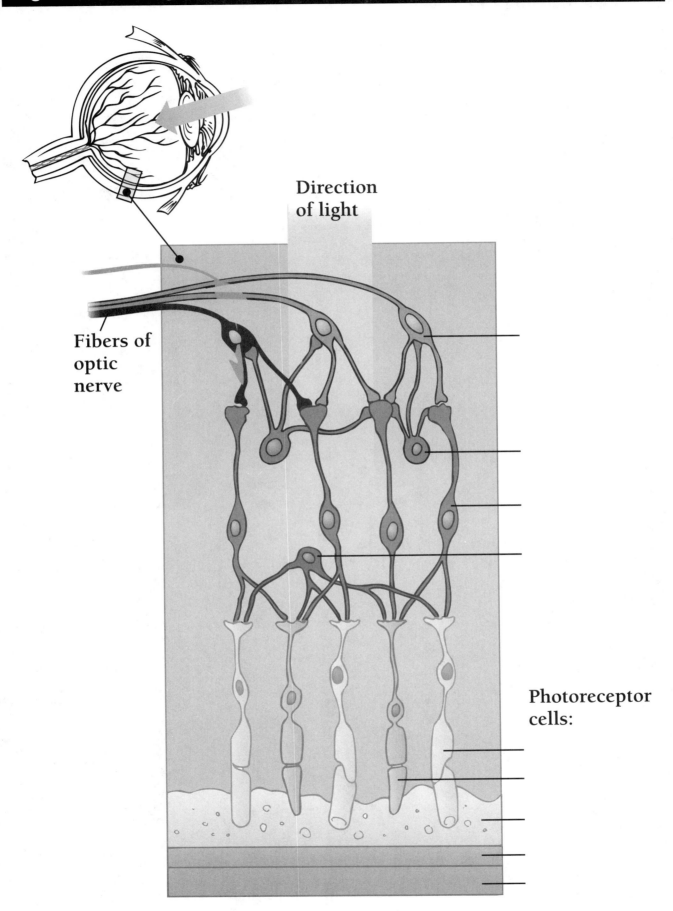

Direction
of light

Fibers of
optic
nerve

Photoreceptor
cells:

Figure 42.14   The Pathway of Visual Information

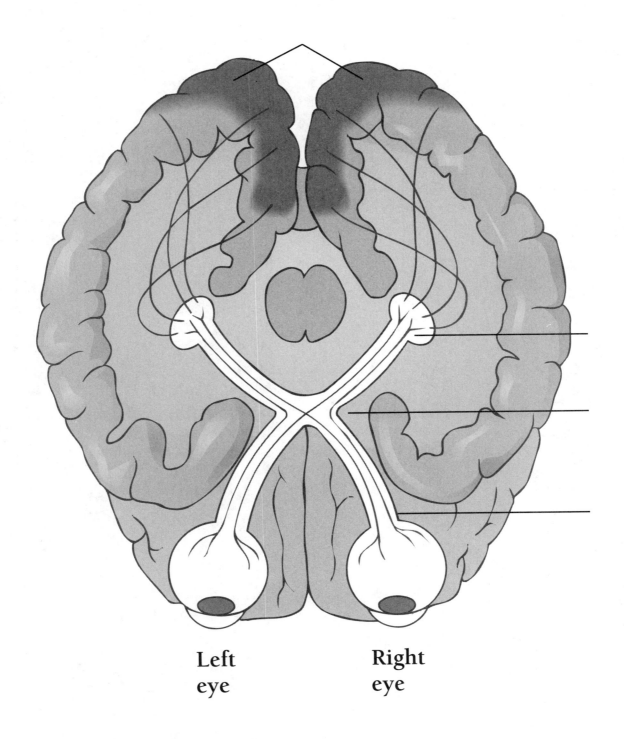

Left
eye

Right
eye

Figure 42.17 Taste Buds

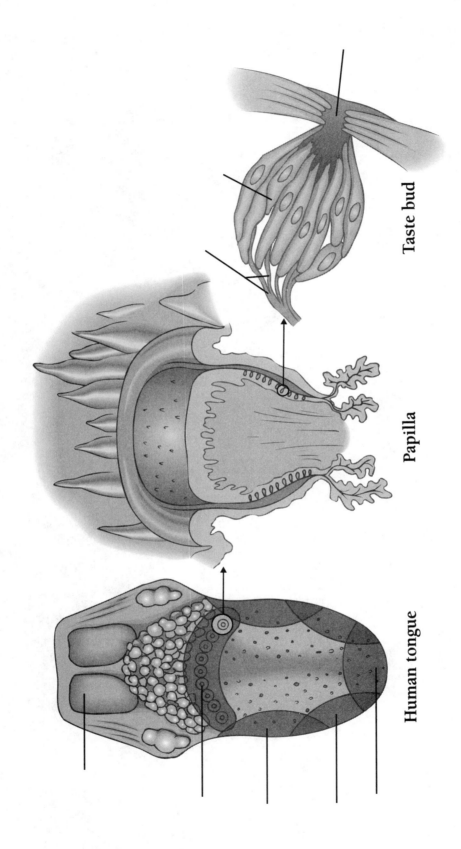

Taste bud

Papilla

Human tongue

Figure 42.18   Olfactory Receptors on the Olfactory Bulb

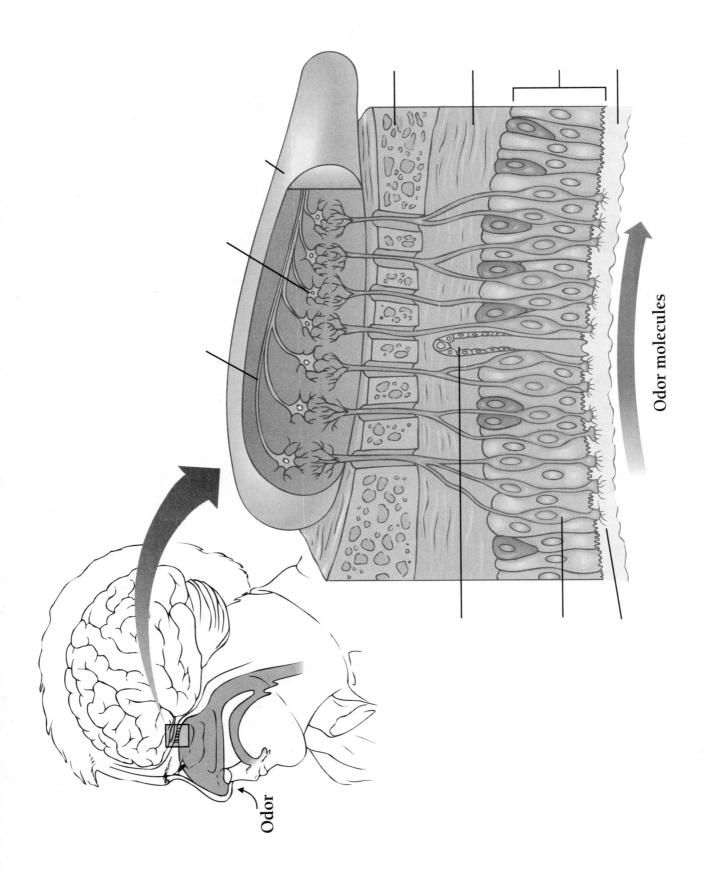

Odor molecules

Odor

Figure 43.3   Female Reproductive Tract in Humans

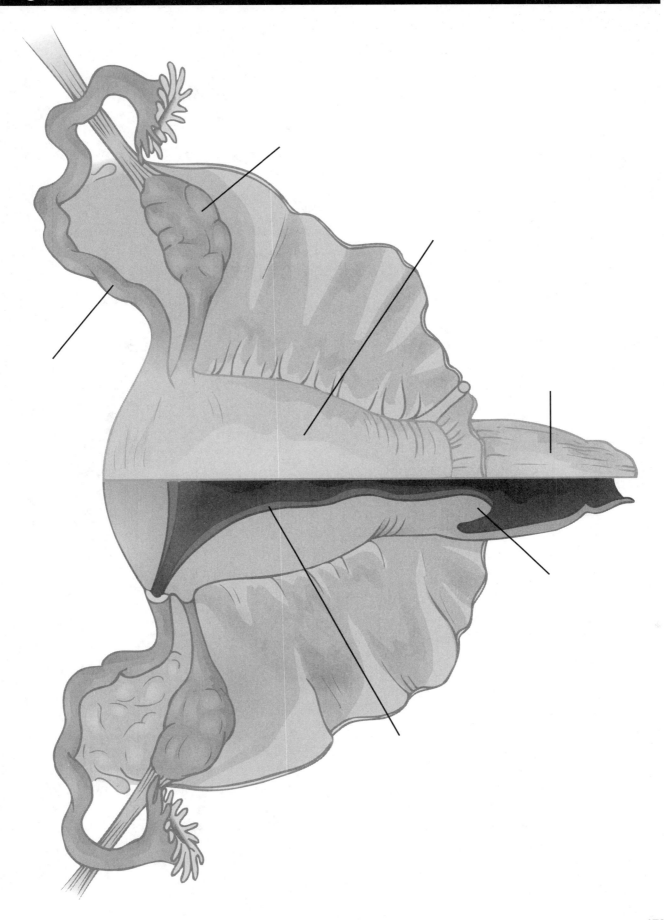

Figure 43.6 The Male Reproductive Tract in Humans

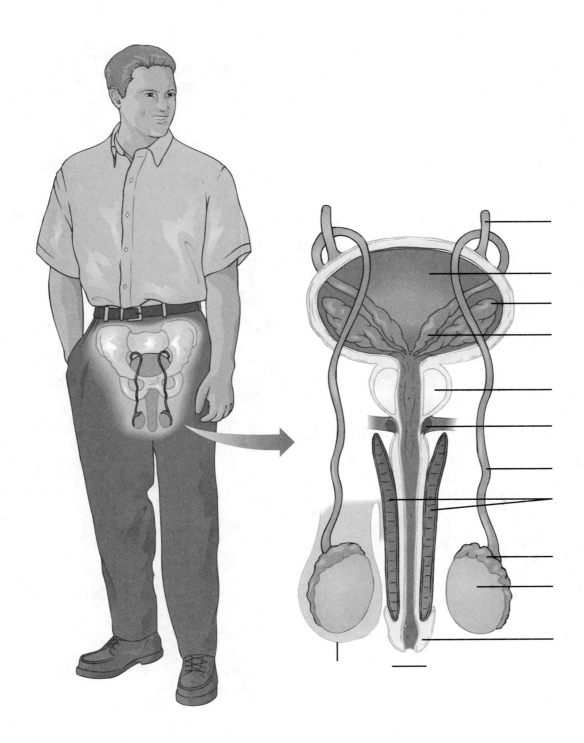

Figure 43.8   Human Sperm

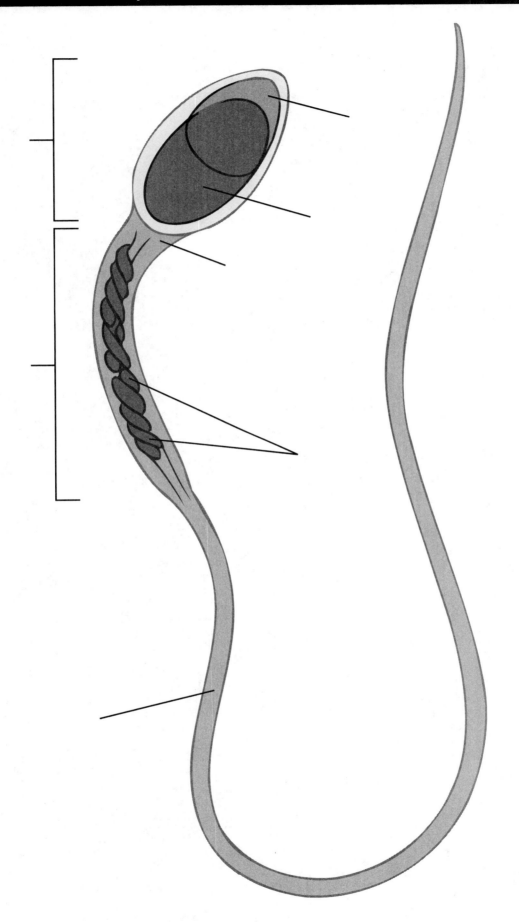

Figure 43.9   Side View of the Male Reproductive Tract

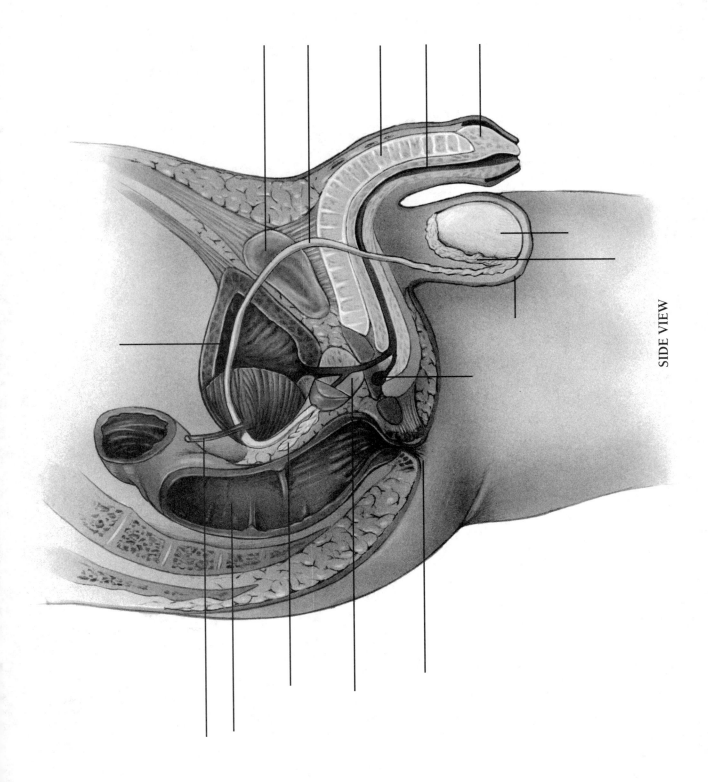

SIDE VIEW

Figure 43.10   Side View of the Female Reproductive Tract

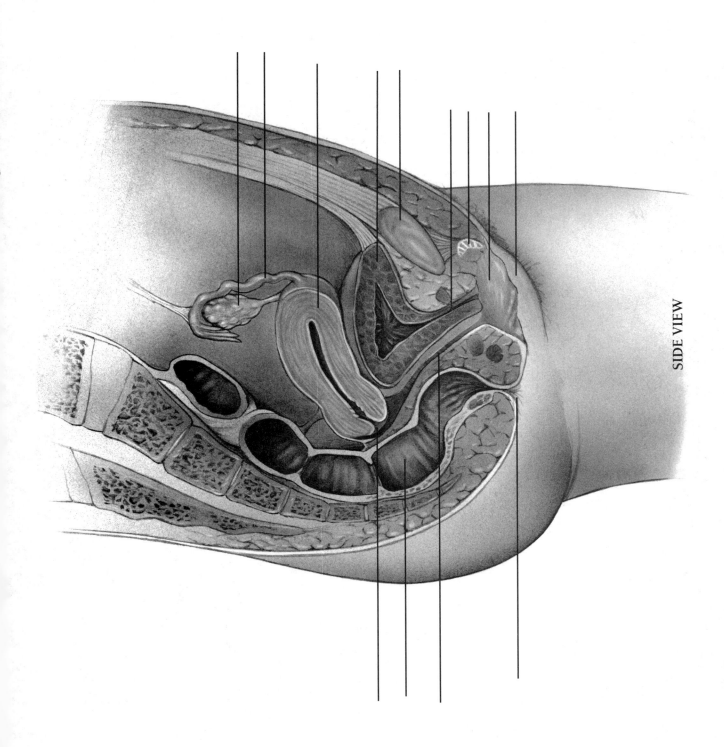

SIDE VIEW

Figure 44.6a Gastrulation in a Sea Urchin

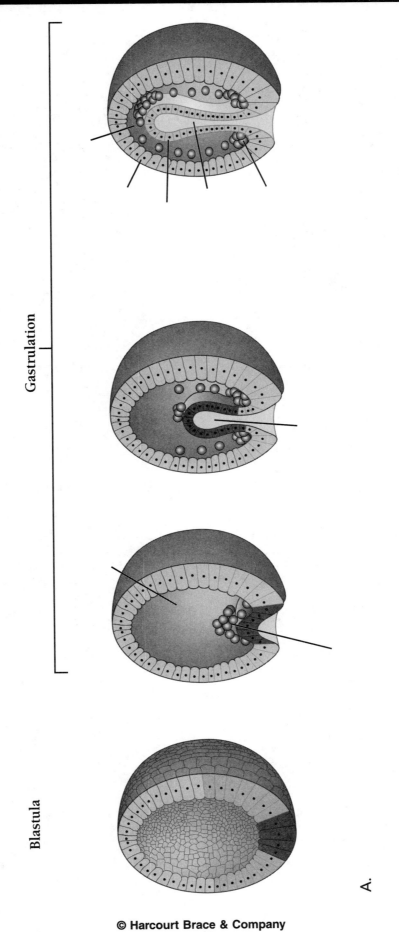

Gastrulation

Blastula

A.

Figure 44.9   A Developing Human

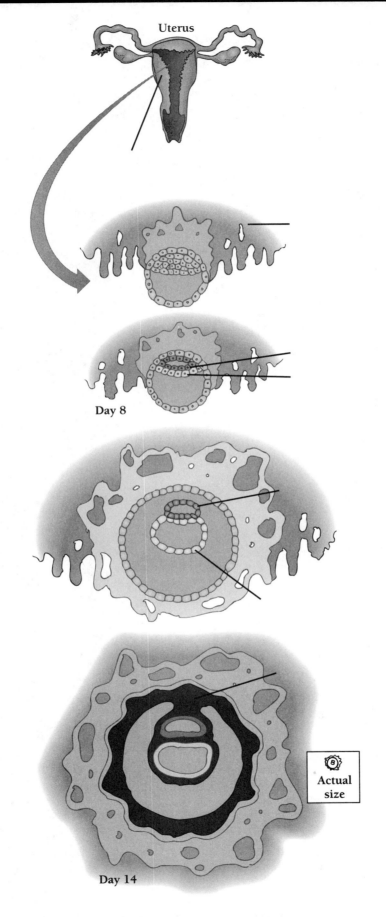

Uterus

Day 8

Day 14

Actual
size

Figure 44.10   Gastrulation in Mammals

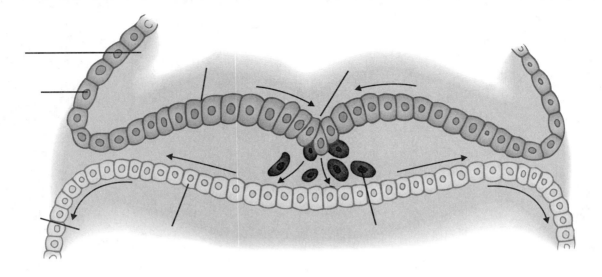

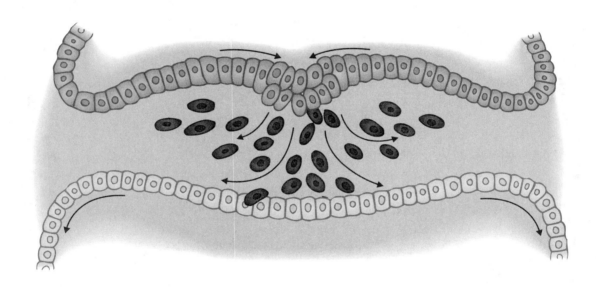

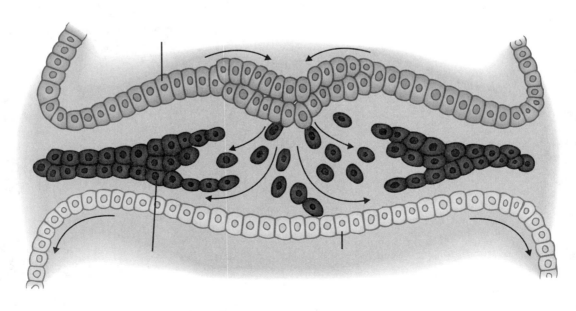

Figure 44.11 Formation of the Neural Tube

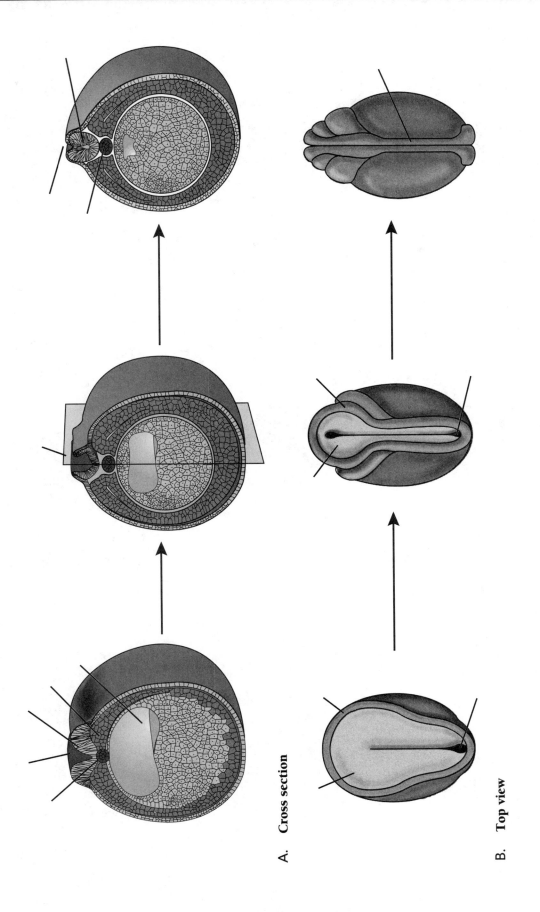

A. Cross section

B. Top view

Figure 44.12   Formation of the Eye

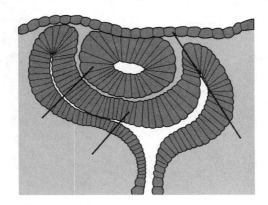

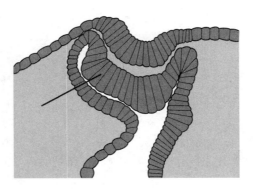

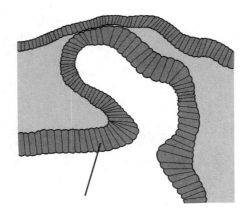

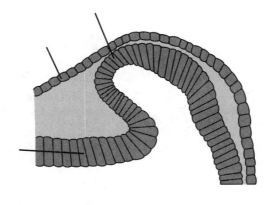